Environmental Assessment in Developing and Transitional Countries

ONE WEEK

Environmental Assessment in Developing and Transitional Countries

Principles, Methods and Practice

Edited by

Norman Lee and Clive George

University of Manchester, UK

JOHN WILEY & SONS, LTD

Chichester · New York · Weinheim · Brisbane · Singapore · Toronto

Other Wiley Editorial Offices

John Wiley & Sons, Inc., 605 Third Avenue,
New York, NY 10158-0012, USA

WILEY-VCH Verlag GmbH, Pappelallee 3,
D-69469 Weinheim, Germany

Jacaranda Wiley Ltd, 33 Park Road, Milton,
Queensland 4064, Australia

John Wiley & Sons (Asia) Pte Ltd, 2 Clementi Loop #02-01,
Jin Xing Distripark, Singapore 129809

John Wiley & Sons (Canada) Ltd, 22 Worcester Road,
Rexdale, Ontario M9W 1L1, Canada

British Library Cataloguing in Publication Data

A catalogue record for this book is available from the British Library

ISBN 0-471-98556-2 (ppc)
ISBN 0-471-98557-0 (paper)

Typeset in 10/12pt Times from the author's disks by Dorwyn Ltd, Rowlands Castle, Hants
Printed and bound in Great Britain by TJ International Ltd, Padstow, Cornwall

This book is printed on acid-free paper responsibly manufactured from sustainable forestry,
in which at least two trees are planted for each one used for paper production.

Contents

Preface

Environmental assessment (EA) originated, as a formalized system of environmental appraisal, at the beginning of the 1970s. Initially, it developed slowly but, from the middle of the 1980s, it has spread to cover most of the developed world. On the whole, EA has been introduced later, and is less firmly embedded, in the development process in low and middle income countries. However, since the late 1980s and especially since the Rio Earth Summit in 1992, there has been a substantial extension in mandatory and other EA procedures in developing countries (otherwise referred to as less developed countries, LDCs) and countries in transition (CITs).

The total volume of EA publications has increased rapidly in response to these developments. However, only a small proportion of the general texts which have been published relate to EA regulations and practice in LDCs and CITs. More fundamentally, few publications reflect the underlying conditions (political, institutional, cultural, economic and environmental) in these countries which shape their EA regulations and practice.

This book aims to remedy this deficiency, and both its contents and contributors have been selected with this in mind. Each author has experience of research, training or practical application of EA in low and middle income countries. Collectively, they cover all of the major regions of the world.

The book is intended for use by *students* and *practitioners*. It is written at a level appropriate to intermediate and advanced levels of study. As well as for EA specialists, it is primarily intended for use in courses relating to environmental planning and management, development studies and project appraisal. It is inter-disciplinary in nature and is accessible from different discipline backgrounds (environmental and other natural sciences, environmental technology, social sciences and development studies and various engineering disciplines). It combines regulatory and procedural reviews, analyses and evaluations of practice, proposals for improvements, and case study examples drawn from a wide variety of countries. It also contains advice on supplementary reading and discussion questions at the end of most chapters, to stimulate and guide further study.

The practitioners for whom the book is intended are of three types. There are those engaged in government administration and non-governmental organizations in low and middle income countries, who are dealing with environmental and development problems. There are those in bilateral aid agencies, international and regional development banks, and other international organizations who have environmental interests and responsibilities relating to LDCs and CITs. Finally there are those in consultancy organizations, universities and technical institutes, whose specialist scientific and technical skills contribute to the EA process.

The book begins with a general introduction (Chapter 1), which provides an overview of the EA process in developing countries and countries in transition, and its role in promoting sustainable development. The remainder of the book is divided into two main parts.

Part 1: EA Principles, Processes and Practice. This contains 10 chapters which first review the economic, environmental and regulatory context in which EA systems operate in LDCs and CITs (Chapter 2), and then overview EA procedures and practice in these countries in six different regions of the world (Chapter 3). Then, procedures and practice are examined in greater detail relating to key activities in the EA process:

- Screening and scoping (Chapter 4)
- Impact prediction and evaluation (Chapter 5)
- Economic valuation of environmental impacts (Chapter 6)
- Social impact assessment (Chapter 7)
- Reviewing the quality of impact assessments (Chapter 8)
- Public and stakeholder participation (Chapter 9)
- Integrated appraisals and decision-making (Chapter 10)
- Environmental monitoring, management and auditing (Chapter 11)

Part 2: Country and Institutional Studies of EA Procedures and Practice. This contains 10 studies prepared by contributors from the countries or institutions concerned. Chapters 12 and 13 contain six country studies, each drawn from a different region in the world. Chapter 12 covers three countries – Chile, Indonesia and Russia – each of which has a number of years of EA experience and whose systems are, in certain respects, relatively developed. In contrast, Chapter 13 covers three countries – Nepal, Jordan and Zimbabwe – whose EIA systems and experience are, in some respects, less developed. Chapter 14 contains three institutional studies – relating to the World Bank, the Asian Development Bank and bilateral aid agencies. All of these review EA procedures and practice and consider ways in which they might be improved.

The book concludes with an *International Perspective on EA Practice in LDCs and CITs* (Chapter 15). It identifies a number of challenges to be addressed, additional measures to be taken if EA is to become a more effective tool for sustainable development, and discusses the rôle of UNEP and other international stakeholders in helping to achieve this.

The completion of this book leaves us indebted to many people. First of all we wish to thank our contributors, who are drawn from many different parts of the world, without whose participation this kind of book would not have been possible. We are grateful to them for sharing their knowledge and experience with us. Each has been encouraged to write in a personal capacity. The views expressed are their own and are not necessarily shared by other contributors or by the organizations to which they are affiliated.

We also thank all those who assisted the preparation of the book by providing information and comments to authors of individual chapters. These are too numerous to mention all of them individually but, on behalf of the contributors, we warmly acknowledge their assistance. Additionally, we thank our colleagues in the EIA Centre, University of Manchester, for their assistance in obtaining documents, providing comments and generally supporting the preparation of the book.

We are also indebted to the publishers and authors of publications who have granted copyright permission to reproduce extracts from their work for inclusion in this book.

Finally, we thank Audrey Lee for her help in the preparation and editing of various drafts of the manuscript of the book, and both Audrey Lee and Kay George for their patience and generous support.

Norman Lee
Clive George
June 1999

List of Contributors

Hussein Abaza   Chief, Economics and Trade Unit, United Nations Environment Programme, Geneva, Switzerland

Ron Bisset   Director, CORDaH Environmental Management Consultants, Edinburgh, UK

Shem Chaibva   Africa Regional Co-ordinator, International Council for Local Environmental Initiatives, Harare, Zimbabwe

Aleg Cherp   Director, Ecologia, Moscow, Russia

Clive George   Senior Research Fellow, EIA Centre, University of Manchester, UK

Luis C. Contreras   Geotechnica Consultants, Santiago, Chile

Ram B. Khadka   Dean, School of Environmental Management and Sustainable Development, Kathmandu, Nepal

Mahmoud Al-Khoshman   Environmental Specialist, Islamic Development Bank, Jeddah, Saudi Arabia

Colin Kirkpatrick   Professor of Development Economics and Director of the Institute for Development Policy and Management, University of Manchester, UK

Norman Lee   Senior Research Fellow, EIA Centre and the Institute for Development Policy and Management, University of Manchester, UK

Bindu Lohani   Manager, Environment Division, Asian Development Bank, Manila, the Phillipines

Remy Paris   Strategic Management of Development Co-operation Division, Development Co-operation Directorate, OECD, Paris, France

Colin Rees   Advisor, Environment Department, World Bank, Washington DC, USA

Batu K. Uprety   Assistant Planning Officer, Ministry of Population and Environment, Government of Nepal, Kathmandu, Nepal

Frank Vanclay   Associate Director, Centre for Rural Social Research, Charles Sturt University, Wagga Wagga, Australia

Christopher Wood   Professor of Environmental Planning and Director of the EIA Centre, University of Manchester, UK

Zulhasni   Co-ordinator, Environmentally Sound Beaches Program, Environmental Impact Management Agency, Indonesia

Abbreviations

ADB	Asian Development Bank
AMDAL	Assessment of environmental impacts (Indonesia)
ARA	Aqaba Region Authority
BAPEDAL	Environmental Impact Management Agency (Indonesia)
BAT	Best available techniques
BKPM	National Investment Board (Indonesia)
BPN	National Land Agency (Indonesia)
CBA	Cost-benefit analysis
CBO	Community-based organization
CEC	Commission of the European Communities
CEQ	Council on Environmental Quality
CIS	Commonwealth of Independent States
CIT	Countries in transition
CONAMA	National Commission for the Environment (Chile)
COREMA	Regional Commission for the Environment (Chile)
CPP	Consultation and public participation
CVM	Contingent valuation method
DAC	Development Assistance Committee
DMC	Developing member countries
EA	Environmental assessment
EBRD	European Bank for Reconstruction and Development
EEAA	Egyptian Environmental Affairs Agency
EFL	Environment Framework Law
EIA	Environmental impact assessment
EIB	European Investment Bank
EID	Environmental impacts declaration
EIS	Environmental impact statement
EMP	Environmental management plan
EMS	Environmental management system
ENVD	Environment Division
ERCD	Environmental and Resource Conservation Division
ETEU	Economics, Trade and Environment Unit
EU	European Union

FMEA	Failure mode effects analysis
GAEAP	Gulf of Aqaba Environmental Action Plan
GCEP	General Corporation for Environmental Protection
GNP	Gross national product
HIA	Health impact assessment
IAP	Interested and affected people
ICES	International Centre for Educational Systems
IEE	Initial environmental examination
IMF	International Monetary Fund
IRDB	InterAmerican Reconstruction and Development Bank
ISA	Initial social assessment
ISO	International Standards Organization
IUCN	World Conservation Union, International Union for Conservation of Nature and Natural Resources
LDC	Less developed countries
MCA	Multi-criteria analysis
MET	Ministry of Environment and Tourism
MOPE	Ministry of Population and Environment
NCS	National conservation strategy
NEAP	National environmental action plan
NEPA	National Environmental Policy Act
NGO	Non-governmental organization
NORAD	Norwegian Agency for Development Co-operation
NPC	National Planning Commission
NPV	Net present value
OECD	Organization for Economic Co-operation and Development
OESD	Office of Environment and Social Development
OP	Operational policy
OVOS	Assessment of impacts on the environment (Russia)
PER	Public environmental review, public environmental expert review
PPA	Pollution action plan
PPP	Policies, plans and programmes
RCA	Environmental qualification resolution (Chile)
REA	Regional environmental assessment
RKL	Environmental management plan (Indonesia)
RPL	Environmental monitoring plan (Indonesia)
SAR	Staff appraisal report
SEA	Strategic environmental assessment, sectoral environmental assessment (World Bank)
SEIA	Summary environmental impact assessment
SEMDAL	Environmental impacts review (Indonesia)
SER	State environmental review, state environmental expert review
SERD	State Environmental Review Department
SIA	Social impact assessment
SIEE	Summary initial environmental examination
TANAPA	Tanzanian National Parks Authority

TEV	Total economic value
TOR	Terms of reference
UNCED	United Nations Conference on Environment and Development
UNDP	United Nations Development Programme
UNECE	United Nations Economic Commission for Europe
UNEP	United Nations Environment Programme
USAID	United States Agency for International Development
USEPA	United States Environmental Protection Agency
WHO	World Health Organization
WTA	Willing to accept
WTP	Willing to pay
WWF	World Wide Fund for Nature

List of Boxes, Figures, Maps and Tables

Boxes

Figures and Map

Tables

1

Introduction

Norman Lee and *Clive George*

1.1 Environmental Assessment, Developing Countries and Countries in Transition

Environmental assessment (EA) is a widely used policy tool for reducing the negative environmental consequences of development activities and for promoting sustainable development. It covers both the assessment of individual development projects, often known as environmental impact assessment (EIA), and the appraisal of development policies, plans and programmes, which is generally referred to as strategic environmental assessment (SEA). In both cases, the purposes of the assessment are:

a) To identify any potentially adverse environmental consequences of a development action, so that they may be avoided, reduced or otherwise taken into account during planning and design
b) To ensure that any such consequences are taken into account, both whilst planning and designing an action and when it is authorized
c) To influence how it is subsequently managed during its implementation.

EA is potentially applicable to any type of development action, which may result in significant environmental impacts, in any part of the world. Its underlying principles are general but the circumstances in which it is applied and, therefore, the particular forms it takes, vary considerably between different parts of the world.

The main focus of this book is upon developing countries (or less developed countries, LDCs) in Africa, Asia, the Middle East and Latin America and countries in transition (CITs) in Central and Eastern Europe and Central Asia. The

Environmental Assessment in Developing and Transitional Countries. Edited by N. Lee and C. George.
© 2000 John Wiley & Sons, Ltd.

individual countries to be covered have been identified according to the level of their per capita national income, using official estimates of Gross National Product (GNP) in 1995 converted to US\$ according to official exchange rates.

The World Bank (1997a) classifies 133 countries into four income categories as follows:

1. Low income economies with annual income per capita of less than 766 US\$. This category includes 49 countries and their per capita incomes range between 80 (Mozambique) and 730 US\$ (Armenia).
2. Lower middle income economies with annual income per capita between 766 and 3160 US\$. This category includes 41 countries and their per capita incomes range between 766 (Lesotho) and 3020 US\$ (Venezuela).
3. Upper middle income economies with annual per capita incomes between 3160 and 9386 US\$. This category includes 17 countries and their average income levels range between 3160 (South Africa) and 8210 US\$ (Greece).
4. High income economies with annual per capita incomes of 9386 US\$ and above. This category includes 26 countries and their average incomes range between 9700 (Republic of Korea) and 40 630 US\$ (Switzerland).

This study relates to the low and middle income countries in the first three categories. Taken together, they occupy 76% of the world's land area and contain 93% of its total population, but only account for 19% of the total GNP of the 133 countries covered.

The geographic distribution of countries, according to income group, is shown in Map 1.1. This shows that:

- Low income countries are mainly located in sub-Saharan Africa, South and East Asia (including China and India) and some parts of Central Asia
- Lower middle income countries are located in parts of the former USSR and some adjoining European countries, the Middle East and North Africa, and Central/South America
- Upper middle income countries are located in parts of Central Europe, the Middle East, South East Asia, South Africa and Central/South America
- High income countries are mainly located in North America, North and Western Europe, parts of South East Asia, Japan, Australia and New Zealand.

The developmental, environmental and regulatory characteristics of low and middle income countries are examined, in their regional context, in Chapter 2. This demonstrates the great variations in many of the characteristics which may influence EA regulation and practice, both *between* low and middle income countries and high income countries, and *among* low and middle income countries themselves. The variability in EA regulations and practice between countries, which often reflects more fundamental differences in their economic, social, political and environmental circumstances, is a recurring theme in this book. In this connection, it is important to emphasise the dangers of indiscriminately transposing conceptions of good EA practice formulated in high income countries to the quite different situations which prevail in many LDCs and CITs.

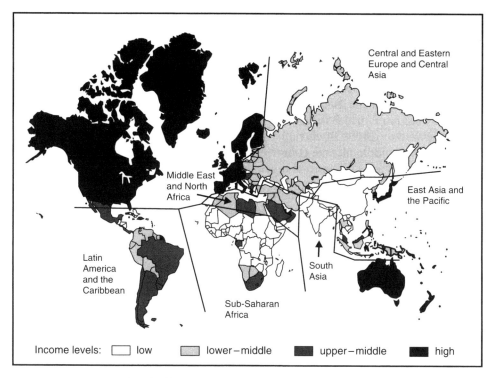

Map 1.1   *Countries Classified by Income Category (1995)*

1.2   The Origins of Environmental Assessment

EA, as a mandatory regulatory procedure, originated in the early 1970s, with the implementation of the National Environmental Policy Act (NEPA) 1969 in the USA. Much of the initial phase of its subsequent development was in a small number of high income countries, such as Canada and Australia, but some developing countries also adopted it at a relatively early stage. Colombia introduced EIA procedures in 1974, and the Philippines established it by presidential decree in 1978 (Smith and van der Wansem 1995).

The major period of expansion of project-level EA has taken place since the mid-1980s. Virtually all high income countries now possess their own mandatory EIA procedures (Lee 1995), as do a large and rapidly increasing number of low and middle income countries. Additionally, most international and bilateral aid agencies and development banks have adopted their own EIA procedures, which they apply when providing development assistance (OECD 1996).

Certain of the earliest regulations (such as NEPA in the USA) covered policy and programming initiatives as well as projects (but they were much less frequently applied to these in practice). In general, SEA regulations and practices have developed at a much slower pace than EIA requirements. However, during the 1990s, both mandatory and less formalized requirements for SEA have been expanding more rapidly not only in high income, but also among lower and middle income countries (Lee 1995; Sadler and Verheem 1996; Therivel and Partidario 1996). This has been mirrored in some strengthening of SEA procedures and practices within aid agencies and development banks. The World Bank, for example, has introduced guidance for both sectoral and regional EA, covering the assessment of plans or programmes which the Bank funds for a specific sector of the economy or geographical region (World Bank 1997b).

A comparative review of current EIA and SEA regulatory provisions and practice in low and middle income countries is presented in Chapter 3, where a number of these features are examined in more detail.

1.3   The Environmental Assessment Process: Scope and Stages

Scope of EA

The types of impacts which are addressed by environmental assessment cover all aspects of the human environment, as well as the ecological and physical environment. A typical assessment might include impacts on:

- Human beings
- Flora and fauna
- Land (including natural resources), water, air and climate
- Cultural heritage assets (including buildings and other structures)
- Landscape and townscape
- Noise and vibration levels
- Eco-systems and other interactions between different components of the environment

This breadth of coverage of the environment in environmental assessment entails overlaps with other forms of impact assessment, including social impact assessment, health impact assessment, risk assessment and cost-benefit analysis. The relationships between environmental assessment and other forms of assessment are explored more fully in Chapters 5–7 and 10.

The projects, to which EIA is applied, may be new developments or major modifications to existing facilities and can occur in a wide range of economic sectors. These include: agriculture, forestry and fishing; mining and other extractive industries; all parts of the energy sector, including fossil-fuelled electricity generation, hydropower, nuclear power and wind power; all major industries within the manufacturing and process industry sector; transport; tourism and leisure developments; water supply; waste treatment and waste disposal facilities; and other infrastructure and urban development projects.

The policies, plans and programmes (PPPs) to which SEA is applied are also wide-ranging and may relate to:

- The overall development of key sectors in the economy (e.g. transport, energy, mining, water supply, forestry and tourism)

- Associated infrastructure development plans, including waste water and solid waste treatment and disposal plans
- Land use and territorial development plans
- National, multi-sectoral PPPs (e.g. privatization programmes and fiscal reform policy measures)
- International and multi-national policy and programme initiatives (e.g. international trade agreements, internationally financed structural adjustment programmes and overseas aid programmes)

Some SEAs may benefit from being co-ordinated with each other and with certain project-level EIAs, within a tiered system of environmental assessment, as illustrated in Figure 1.1. In such a system, a more strategic form of environmental assessment is first applied to selected policies, plans and programmes during the early stages of the development planning cycle. Then, as shown in Figure 1.1, account is taken of these higher level assessments at each subsequent level in the planning structure. In this way, SEA and EIA are intended to be complementary to each other; each performs tasks most appropriate to the phase of the development planning process at which it is to be used. In practice, as discussed in subsequent chapters, tiering arrangements are still in their infancy in many countries.

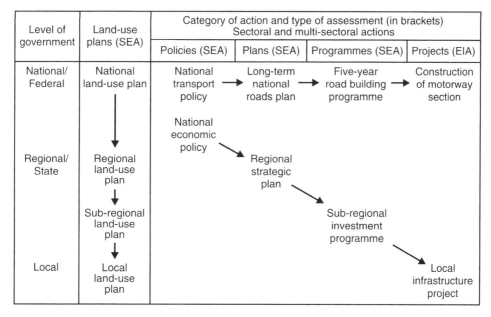

Figure 1.1 *Sequence of Actions and Assessments Within a Tiered Planning and Assessment System.* Source: Lee and Walsh (1992)

Stages in the EA Process

SEA and EIA processes, although applied during different phases of the planning and project cycle, contain similar kinds of assessment activities. These may include:

- *Screening*: deciding whether the nature of the proposed action and its likely impacts are such that it should be submitted to environmental assessment
- *Consideration of alternatives*: a review of alternatives to the proposed action (policy, plan, programme or project)
- *Description of the action*: describing the action in a suitable form to enable its effects to be predicted
- *Description of the environmental baseline*: describing the current state of the environment likely to be affected, and its expected future state in the absence of the proposed action
- *Impact identification and scoping*: determining which environmental impacts should be investigated in the assessment
- *Prediction of impact magnitude and significance*: determining how large the impacts are likely to be, and assessing their importance
- *Identification of mitigation measures*: defining what steps can be taken to eliminate or reduce any significant impacts or to compensate for them
- *Preparing the documentation of the assessment*: documenting the findings of the assessment (for example, in an environmental impact statement) in a manner that is clearly understandable to those involved in consultations and decision making
- *Review*: evaluating the documentation to determine its adequacy for consultation and decision-making purposes
- *Consultation and public participation*: enabling the environmental authorities and the public to comment upon the proposed action and its environmental impacts, based upon the documentation of the assessment (N.B. consultation and public participation may also take place at other stages of the process, notably in scoping)
- *Decision-making*: using the assessment documentation and consultation findings to reach a decision on the authorization of the proposed action, with or without conditions attached
- *Monitoring implementation*: checking whether the action is implemented in accordance with any environmental conditions of the decision and whether its environmental performance is consistent with the assessment's predictions

These stages and activities in the EA process, and the assessment methods used within them, are reviewed more fully in Chapters 4–11.

1.4   Environmental Assessment and Sustainable Development

The overall purpose of EIA and SEA is to assist in shaping the development process, not to prevent development from taking place. More precisely, their role is to ensure that the environmental consequences of development proposals are systematically assessed and taken into account, in conjunction with their likely economic, social and other consequences, when determining development strategies and, later, when approving individual development projects.

Other forms of appraisal, such as cost–benefit analysis (CBA) and social impact assessment (SIA), may be used to assess the economic and social consequences of developments, so that they can be taken into consideration alongside the findings of the environmental assessment (Vanclay and Bronstein 1995; Kirkpatrick and Lee 1997). However, as discussed in Chapter 10, the integration of these different forms of appraisal, and their combined use for decision-making purposes, can be quite complex from both a procedural and a methodological standpoint. Yet, the pressure for integrated appraisal (sometimes called sustainability appraisal) grows as political commitments to sustainable development increase.

The 1992 United Nations Conference on Environment and Development (UNCED) gave considerable impetus to the adoption, by international organizations and national governments, of sustainable development objectives. It also recognized the role of environmental assessment in their attainment (see Box 1.1). This has two important consequences for EA procedure and practice:

- It reinforces existing tendencies to improve procedures and methodologies for more integrated forms of appraisal and decision-making in the development process
- It highlights the need to develop methods for assessing the significance of environmental, economic and social impacts according to sustainable development criteria

These consequences require that the term 'sustainable development' be given sufficient operational meaning. This was explicitly recognized at UNCED, in Agenda

---

Box 1.1   Extracts from the Rio Declaration on Environment and Development and Agenda 21

Principle 17 of the Rio Declaration

Environmental impact assessment*, as a national instrument, shall be undertaken for proposed activities that are likely to have significant adverse impact on the environment and are subject to a decision of a competent national authority.

Agenda 21 proposes that governments should:

- Promote the development of appropriate methodologies for making integrated energy, environment and economic policy decisions for sustainable development, inter alia, through **environmental impact assessments** (9.12(b))
- Develop, improve and apply **environmental impact assessments** to foster sustainable industrial development (9.18)
- Carry out investment analysis and feasibility studies, including **environmental impact assessment**, for establishing forest-based processing enterprises (11.23(b))
- Introduce appropriate **environmental impact assessment** procedures for proposed projects likely to have significant impacts upon biological diversity, providing for suitable information to be made widely available and for public participation, where appropriate, and encourage the **assessments of the impacts of relevant policies and programmes** on biological diversity (15.5(k))

Sources: UNCED Report A/CONF.151/5/Rev.1 13, June 1992, Agenda 21 14 June 1992.

* The UNCED reports use the term 'environmental impact assessment' to include both strategic-level and project-level environmental assessments.

21's proposals for the development of national and global indicators of sustainable development.

The phrase sustainable development first came to notice in the *World Conservation Strategy: Living Resource Conservation for Sustainable Development*, published jointly by the World Conservation Union (IUCN), the United Nations Environment Program (UNEP) and the World Wide Fund for Nature (WWF) in 1980. It became more widely known through the publication in 1987 of *Our Common Future*, the report of the World Commission on Environment and Development chaired by Gro Harlem Brundtland. The phrase has subsequently secured wide international recognition following UNCED, the Rio earth summit of 1992.

The most widely used definition of sustainable development is still that derived from the Brundtland report, as:

> [development which] meets the needs of the present without compromising the ability of future generations to meet their own needs.

An alternative definition was put forward in the IUCN/UNEP/WWF report *Caring for the Earth* in 1991, as:

> improving the quality of human life while living within the carrying capacity of supporting ecosystems.

The Rio Declaration (Principle 3) added a further dimension to the definition:

> to equitably meet developmental and environmental needs of present and future generations.

Three guiding principles may be derived which assist in the formulation of criteria and indicators for use in environmental assessments and other appraisals:

- *Intergenerational equity*. Underlying this principle is the notion of passing on an equivalent resource endowment to the next generation, so that it has at least an equal opportunity to meet its needs as the present generation
- *Intragenerational equity*. In addition to Principle 3's call for equity, Principle 5 of the Rio Declaration requires that 'all states and all people shall co-operate in the essential task of eradicating poverty as an indispensable requirement for sustainable development, in order to decrease the disparities in standards of living and better meet the needs of the majority of the people of the world'
- *Carrying capacity*. In this case the guiding principle is that the ability of an ecosystem to support life is limited ultimately by the system's capacity to renew itself or safely to absorb wastes

These principles provide some guidance on the ways in which impact assessments may be strengthened to incorporate sustainable development criteria. These include:

- Greater attention to predicting and evaluating the impact of developments on natural resource stocks, and on total national capital

- Greater use of strategic environmental assessment to assess the medium, longer term and cumulative impacts of developments, in relation to ecosystem security and waste absorption capacity
- More explicit consideration of the economic and social implications of developments, particularly for the poorer, disadvantaged sections of communities, either through separate economic and social appraisals or within more integrated forms of impact assessment
- A greater focus on identifying impacts which may be irreversible
- A greater combined use of integrated appraisals and stakeholder involvement in decision-making processes

The International Study of the Effectiveness of Environmental Assessment (Sadler 1996) drew attention to approaches such as these, and to the need for EA to incorporate them at the level of global as well as national and local impacts. At the global level, this entails taking account of the differing past and present consumption and pollution loads of countries at different stages of development, and of relevant international conventions and other agreements.

The ways in which environmental assessments, and other forms of appraisal, may be adapted to serve sustainable development objectives is a challenge facing all countries and organizations. It is a theme which is revisited in a number of the following chapters.

Benefits and Costs of EA Systems

The value of environmental assessment as an appraisal tool depends, in the final analysis, on the relationship between the benefits and costs of its application as illustrated in Box 1.2.

Box 1.2   Benefits and Costs of EA Systems

Benefits

1. Environmental and other sustainability benefits, attributable to the EA system, resulting from modifications to actions prior to their approval and implementation.
2. Savings in the mitigation costs due to earlier detection of potential environmental problems and better designed corrective measures to deal with these problems.
3. Savings in time in obtaining approval for new developments, also due to the earlier detection and correction of environmental problems which reduce controversy and conflict during the authorization process.

Costs

1. Extra costs to the developer and the authorities in complying with EA study and procedural requirements.
2. Losses of time where the system does not work efficiently and unjustified delays occur.
3. Additional mitigation expenditures due to the EA process commencing too late in the planning and project cycle or where it is used to impose insufficiently substantiated mitigation requirements on developers.

The quantification of these benefits and costs for individual countries is a difficult task, but the main conclusions to be drawn from the available studies (e.g. European Commission 1996a) are reasonably clear:

- The benefits of well-functioning EA systems usually exceed their costs of implementation; but
- In a number of cases, their full potential is not being realized because either their full benefits are not being achieved, or their costs of implementation are higher, than those which are achievable

These are important findings for low and middle income countries. In particular, they highlight the need to look beyond the progress being made in approving new EA regulations, to the quality of EA practice which is being realized in such countries. In subsequent chapters, the causes of under-performance in practice and the means of addressing these, are examined in greater detail.

The Book's Structure

The remainder of the book is divided into two main parts:

- *Part 1: EA Principles, Processes and Practice (Chapters 2–11)*. This first reviews the economic, environmental and regulatory context in which EA systems operate in LDCs and CITs (Chapter 2) and overviews EA procedures and practices in these types of countries in six different regions of the world (Chapter 3). Then, procedures and practice are examined in greater detail relating to different stages and key activities in the EA process (Chapters 4–11). Each of these chapters includes guidance on further reading and discussion questions to assist further study.
- *Part 2: Country and Institutional Studies of EA Procedures and Practice*. This contains a collection of empirical studies, prepared by contributors from the LDCs, CITs and international institutions to which they relate, which exemplify leading issues identified and analysed in Part 1. Six country studies are included, covering Chile, Indonesia and Russia (Chapter 12) and Nepal, Jordan and Zimbabwe (Chapter 13). Chapter 14 contains studies of EA procedures and practices in the World Bank, the Asian Development Bank and bilateral aid agencies. The book concludes with an international perspective on EA practice in LDCs and CITs (Chapter 15). Measures are proposed to make EA a more effective tool for sustainable development and the rôle of international stakeholders in promoting this is reviewed.

Discussion Questions

1. What are the main similarities and differences between EIA and SEA? What reasons can be suggested for the slower development of SEA than EIA in low and middle income countries?

2. What characteristics of developing countries and of countries in transition may influence their environmental assessment procedures and practices, and how?
3. What contribution can EA make to achieving sustainable development? What are the main limitations in its capacity to promote sustainable development?

Further Reading

A useful introduction to the basic principles of environmental impact assessment in a developing country context is provided in UNEP (1988). A short historical overview of EIA and SEA developments in different parts of the world can be obtained from Lee (1995), supplemented by Sadler (1996), United Nations Environment Programme (1996), Sadler and Verheem (1996) and Bellinger *et al.* (1999). Vanclay and Bronstein (1995) provide useful surveys of different types of environmental, social and economic appraisal and Kirkpatrick and Lee (1997) examine, with case study illustrations, a number of issues relating to the integration of different appraisal methods. Operationalizing the sustainable development concept and constructing sustainable development indicators for use in appraisal and decision-making are discussed in Sadler (1996), and, in relation to global impacts, in George (1999).

References

Bellinger, E, Lee, N, George, C, Paduret, A (eds) (1999) *Environmental Assessment in Countries in Transition*, Central European University Press, Budapest (in press)

European Commission (1996a) *Environmental Impact Assessment: a Study on Costs and Benefits.* DGXI, Brussels

George, C (1999) Testing for sustainable development through environmental assessment: criteria and case studies, *Environmental Impact Assessment Review* **19**: 175–200

IUCN/UNEP/WWF (1980) *World Conservation Strategy: Living Resource Conservation for Sustainable Development* Gland, Switzerland

IUCN/UNEP/WWF (1991) *Caring for the Earth: a Strategy for Sustainable Living* Gland, Switzerland

Kirkpatrick, C and Lee, N (eds) (1997) *Sustainable Development in a Developing World: Integrating Environmental Assessment with Socio-Economic Appraisal*, Edward Elgar, Cheltenham

Lee, N (1995) Environmental assessment in the European Union: a tenth anniversary, *Project Appraisal* **10**: 77–90

Lee, N and Walsh, F (1992) Strategic Environmental Assessment: an overview, *Project Appraisal* **7**: 126–136

Organization for Economic Co-operation and Development (1996) *Coherence in Environmental Assessment: Practical Guidance on Development Co-operation Projects*, OECD, Paris

Sadler, B (1996) *Environmental Assessment in a Changing World: final report of the International Study of the Effectiveness of Environmental Assessment*, Canadian Environmental Assessment Agency, Ottawa

Sadler, B and Verheem, R (1996) *Strategic Environmental Assessment Status, Challenges and Future Directions*, No 53, Ministry of Housing, Spatial Planning and the Environment, The Hague

Smith, DB and van der Wansem, M (1995) *Strengthening EIA Capacity in Asia: Environmental Impact Assessment in the Philippines, Indonesia, Sri Lanka* World Resources Institute, Washington DC

Therivel, R and Partidario, MR (eds) (1996) *The Practice of Strategic Environmental Assessment*, Earthscan, London

United Nations (1992) *Report of the United Nations Conference on Environment and Development*, UNCED Report A/CONF.151/5/Rev.1, 13 June 1992

United Nations Environment Programme (1988) *Environmental Impact Assessment: Basic Procedures for Developing Countries*, UNEP, Bangkok

United Nations Environment Programme (1996) *EIA: Issues, Trends and Practice* (prepared by Bisset, R), UNEP, Nairobi

Vanclay, F and Bronstein, DA (eds) (1995) *Environmental and Social Impact Assessment*, John Wiley, Chichester

World Bank (1997a) *World Development Report 1997: The State in a Changing World*, Oxford University Press, Oxford

World Bank (1997b) *The Impact of Environmental Assessment: The World Bank's Experience* (Second Environmental Assessment Review), World Bank, Washington DC

World Commission on Environment and Development (1987) *Our Common Future*, Oxford University Press, Oxford

PART ONE

EA Principles, Processes and Practice

2

Environmental Assessment in its Developmental and Regulatory Context

Norman Lee

2.1 Introduction

This chapter reviews the developmental and regulatory context in which EA provisions are formulated and implemented in low and middle income countries.

Sections 2.2 and 2.3 use a simple systems approach to examine:

- The relationships between economic development and environmental quality
- The relationships between regulatory systems (of which EA procedures form a part) and the economic, social and environmental systems whose performance they seek to improve

The second part of the chapter (Sections 2.4–2.6) examines how these development–environment–regulatory relationships operate in low and middle income countries in different regions of the world. It reviews:

- The economic and social changes taking place
- The changes in environmental pressures and environmental quality which result
- The characteristics and effectiveness of the regulatory systems (environmental and general) which influence these

---

Environmental Assessment in Developing and Transitional Countries. Edited by N. Lee and C. George.
© 2000 John Wiley & Sons, Ltd.

The picture which emerges is of an underlying similarity in the forces at work in all countries but of considerable differences in their detailed content and intensity between low, middle and high income countries, between different regions and between different countries within the same region.

The concluding section of the chapter (Section 2.7) summarizes the relationships between environmental assessment provisions in low and middle income countries and the environmental regulatory systems and development processes to which they relate. It also makes proposals concerning how EA provisions might be developed in the light of these relationships.

2.2   Relationships Between the Development Process and Environmental Quality

The links between economic and environmental systems are illustrated in a simplified economic-environmental model in Figure 2.1.

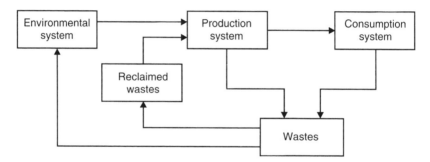

Figure 2.1   *Linkages Between Economic and Environmental Systems*

During the economic development process, natural resources (water, minerals, wood, crops etc.) are extracted from the environmental system, processed into goods and services and distributed to consumers. Residuals (atmospheric emissions, effluent discharges, solid wastes, noise, surplus heat etc.) from extraction, production and consumption activities become wastes. These wastes are either reclaimed and recycled for use in future production or are returned to the environmental system, with or without further treatment.

In this model, development may influence the sustainability of the environmental system in two ways:

- *Through the abstraction of natural resources.* Depending upon the rate of resource abstraction relative to the size of the resource stock, and on whether the resources in question are renewable or non-renewable, development could give rise to a *resource conservation* problem
- *Through the discharge of residuals.* Depending on their nature, scale and location, and on the carrying capacity of the receiving environment, development could give rise to an *environmental pollution* problem

These two influences may damage the environmental system to the point where it can no longer supply certain natural resources or secure certain life-support systems. If so, the sustainability of the development process is threatened and its economic and social goals will remain unfulfilled.

Whether or not development is sustainable depends on four sets of factors:

1. The rate and composition of economic growth.
2. The resource and residual coefficients which, taken into consideration with the rate and composition of growth, determine the size of the resource and residual flows between the economic and environmental systems.
3. The carrying capacity of the environmental system relative to the resource abstraction and residual discharge flows imposed upon it.
4. The responses of society to economic and environmental change.

In turn, these are affected by the interplay between the market forces (domestic and international) impacting on each country and the environmental protection and economic policies which its government decides to apply. *Market forces* are driven by such variables as incomes, tastes, relative prices, foreign trade, market induced technical changes etc. *Economic, social and environmental protection policies* modify these market forces, as well as influencing the relationship between development and environmental quality, as shown below.

2.3   Policy Instruments of Sustainability

Environmental Policy Instruments

The primary purpose of these instruments, which are grouped into three categories below, is to protect or improve environmental quality – but they may also have important economic and social consequences.

Command and control instruments

Typically these take the form of permit and authorization procedures relating to:

- The types of products that may be produced and used
- The types and quantities of raw materials that may be abstracted and used for production and consumption
- The technologies by which goods and materials may be produced
- The maximum quantities and types of residuals which may be released into the environment
- The locations at which resource abstraction, production and other activities may take place

Economic instruments

These influence behaviour to improve environmental performance through the use of:

- Pollution charges and environmental taxes
- Environmental protection subsidies and grants
- Market creation schemes, such as emission trading schemes
- Environmental licensing charges and fines for non-compliance with environmental regulations

Planning and other instruments

These include:

- Environmental planning studies
- Environmental assessment (SEA and EIA) measures
- Environmental audit procedures and environmental management systems
- Voluntary agreements to encourage compliance with environmental quality targets through such measures as industry covenants etc.

Economic and Social Policy Instruments

The primary purpose of these instruments is to serve economic or social objectives – but they may also have important environmental side-effects. They include similar categories of instruments to those for environmental policy instruments but their content and policy orientation are different. They include:

- *Command and control measures* relating, for example, to the ownership and use of land and property, and to monopoly and foreign trade regulation
- *Economic instruments* including a wide range of fiscal measures (e.g. taxes, subsidies, grants and government spending programmes) serving economic and social purposes
- *Planning and other instruments* including the preparation of general, regional and local area development plans, and sectoral development plans (for transport, energy, minerals, tourism, water etc.)

The relationships between environmental and economic/social policy instruments and their environmental and other consequences are illustrated in Figure 2.2. The

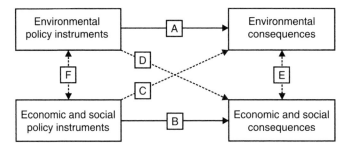

Figure 2.2   *Linkages Between Environmental Policy Instruments and Economic and Social Policy Instruments and the Consequences of their Application*

continuous lines indicate the consequences which are directly connected to the primary purpose of each of the policy instruments. The discontinuous lines identify the other potentially important consequences which may indirectly result.

Environmental assessment, as a policy instrument, fits into this scheme in a number of ways. In combination, these demonstrate its central role in the promotion of sustainable development – and its dependence on related policy instruments to perform that role effectively. For example:

- It is an appraisal instrument within the environmental protection sector. It contributes to the application of environmental licensing and development permitting schemes. It can also assist in the appraisal of government environmental expenditure programmes, the appraisal of economic instruments for environmental protection and environmental planning, life cycle analyses etc.
- It is also an environmental appraisal tool, for use alongside economic and social appraisals, in formulating economic and social policies and other measures relating to: (a) privatization, de-regulation, trade liberalization, structural adjustment programmes; (b) taxation measures and government spending proposals for development and infrastructure programmes; and (c) other development policies, plans and programmes of an economic and social character

At the same time, the scope of application and level of effectiveness of EA instruments depends, to a significant degree, on the scope and effectiveness of the environmental, economic and social policy instruments into which they are introduced, and the degree to which the EA appraisal process is successfully integrated into their procedures and practices. In both these respects, shortcomings are commonly found.

In a supportive cultural, political and institutional context, EA can contribute to the increased effectiveness of other policy instruments through its requirements for more systematic data gathering and analysis, more extensive consultations and greater transparency, and for more careful development planning, decision-making and implementation. On the other hand, where favourable conditions do not exist, EA may intensify existing problems of conflict, delay and unsatisfactory implementation unless other corrective measures are taken.

2.4   Development and Environmental Quality Relationships in Low and Middle Income Countries

The following review of development-environment relationships uses similar regional groupings of countries as in Chapter 1. It covers: Sub-Saharan Africa; East Asia and the Pacific, including China; South Asia; Europe and Central Asia; Middle East and North Africa; and Latin America and the Caribbean. Some data are also provided, relating to high income countries (all regions combined), for comparative purposes.

The information is presented within the framework of a simple 'pressure-state-response' model where:

- One set of variables describes the development pressures on a country's environmental quality and natural resource base. These pressures are assumed to be of two main types: population-related and economic activity-related
- A second set describes the resulting state of the environment and the major concerns associated with this
- A third set describes the principal characteristics of the regulatory systems which exist to deal with these concerns

The first two sets of variables are reviewed in this Section and Section 2.5 and the third is examined in Section 2.6.

Development Pressure Indicators

Population-related variables

Data relating to a number of population variables are presented in Table 2.1. Simple population density measures are only a crude measure of population pressure on land and other resources. Nevertheless, they reveal very considerable disparities between South and East Asia and all other regions, including the high income region. Development pressures change over time according to the rate of population growth. All regions, except the Europe and Central Asia region, record population growth rates which are almost 2–3 times greater than in the high income economies.

The degree of urbanization is currently much higher in the high income countries than in all other regions except the Latin America and Caribbean region. However, the rate of urbanization, measured according to the annual rate of growth in the urban population, is very much higher in the low and middle income economies than in the high income region. Urban population growth rates in Sub-Saharan

Table 2.1   *Population-related development pressures, classified by region*

Region	Population density (persons/sq. km) 1995	Population growth (%, annual) 1980–1990	Population growth (%, annual) 1990–1995	Urban Population (% of total population) 1995	Urban Population growth (%, annual) 1980–1995
Low and middle income					
Sub-Saharan Africa	24.0	2.7	2.6	31	5.0
East Asia and Pacific	105.0	2.3	1.3	31	4.2
South Asia	242.2	2.1	2.1	26	3.4
Europe and Central Asia	20.0	0.6	0.5	65	1.6
Middle East & North Africa	24.7	3.2	3.3	56	4.2
Latin America & Caribbean	23.4	3.0	2.3	74	2.8
All low and middle income	47.0	2.2	1.7	39	3.3
All high income	28.2	1.2	0.9	75	0.7

Source: World Bank 1997.

Africa, for example, were seven times higher than in the developed economies over the previous 15 years.

Economic activity-related variables

Data relating to a number of these variables are presented in Table 2.2. Average incomes in high income countries in 1995 were more than seven times higher than in the richest regional group of low and middle income countries (Latin America and the Caribbean) and more than 70 times greater than in the poorest regional group (South Asia). Furthermore, the disparities between high and low/middle income regional groups have tended to grow over time (except in the case of East and South Asia).

The relationship between income levels and growth rates and pressures on the environment, within individual countries, is complex. First, major disparities in per capita income tend to be reflected in *per capita consumption of different resources*. As shown in Table 2.2, energy consumption per capita in the high income country groups in 1994 was over five times greater than in Latin America and the Caribbean and over 20 times greater than in South Asia. However, resource *extraction*, which is more closely associated with pressure on the natural resource stock in a country,

Table 2.2   *Economic activity-related development pressures, classified by region*

Region	GNP per capita (US$)	GNP per capita growth (%, annual)	Industrial sector growth (%, annual)		Energy use per capita, oil equivalent (kg)	Energy consumption growth (%, annual)
	1995	1985–1995	1980–1990	1990–1995	1994	1980–1994
Low and middle income						
Sub-Saharan Africa	490	−1.1	0.6	0.2	237	1.2
East Asia and Pacific	800	7.2	8.9	15.0	593	4.8
South Asia	350	2.9	6.9	5.3	222	6.4
Europe and Central Asia	2220	−3.5	NA	NA	2647	−4.6
Middle East & North Africa	1780	−0.3	1.1	NA	1200	4.8
Latin America & Caribbean	3320	0.3	1.4	2.5	960	2.9
All low and middle income	1090	0.4	(3.9)	(4.9)	739	−0.1
All high income	24930	1.9	3.2	0.7	5066	1.9

NA = not available. Numbers in brackets denote incomplete coverage.
Source: World Bank 1997.

may follow a different pattern due to the influence of international trade on the rate of resource abstraction.

A second complicating variable relates to differences in industrial structure between individual countries and regional groups. The relative importance of the industrial sector, and of more heavily polluting industries within this, is of particular significance. These proportions were particularly high in a number of countries within the CIT region, prior to the period of transition.

During more recent years, the main growth in the high income countries has been in the services sector and the non-basic industrial sector where development pressures on the environment tend to be less. In contrast, the relative importance of the more highly polluting industries grew considerably, between 1980–1995, in certain low and middle income regions (e.g. East and South Asia and Latin America). If high income countries obtain natural resources and the products of heavily polluting industries through imports, they are net exporters of environmental pressures to other countries. Low and middle income countries, which export primary products and processed materials from heavily polluting industries, are often net importers of environmental pressures from other countries.

A third, complicating factor is that, where income levels and growth rates are higher, mitigation may be given a higher political and spending priority. Whether or not the development process contains or can create sufficient self-correcting mechanisms to secure its own sustainability is an unresolved issue. According to the World Bank, 'some environmental problems decline as income rises; some initially worsen but then improve as incomes rise; other indicators of environmental stress continue to worsen as incomes rise so long as the technical solution is considered too expensive relative to the perceived benefits of its mitigation.' (World Bank 1992, p. 10).

2.5   Environmental Pressures, Trends and Concerns

UNEP has attempted to identify the main environmental trends and concerns in several regions of the world, based on regionally-based consultations (UNEP 1997). Its findings are summarized in Table 2.3 using a regional classification system which differs, in certain respects, from that used in earlier tables within the chapter.[1] They highlight the major areas of environmental concern in each region which, in the views of those consulted, need to be addressed through new or strengthened policy measures. These major concerns are summarized below.

Major Concerns in Individual Regions

Africa

● One-third of all cropland and permanent pasture is modestly or severely degraded

---

[1] Sub-Saharan Africa and Middle East and North Africa in Tables 2.1 and 2.2 are reclassified as Africa and West Asia in Table 2.3. East Asia and Pacific and South Asia are combined into Asia-Pacific in Table 2.3. Europe and Central Asia is re-entitled Europe and CIS countries

Table 2.3   *Regional environmental trends and concerns*

Environmental problem	Africa		Asia-Pacific		West Asia		Latin America & Caribbean		Europe and CIS countries		North America	
	(a)	(b)	(a)	(b)	(a)	(b)	(a)	(b)	(a)	(b)	(a)	(b)
Land degradation	✔	✔	✔	✔	✔	✔	✔	✔				
Forest loss and degradation	✔	✔	✔	✔	✔		✔	✔				
Loss of biodiversity and habitat fragmentation	✔		✔	✔	✔		✔		✔	✔		
Access to fresh water, water pollution	✔	✔	✔	✔	✔	✔	✔			✔		✔
Degradation in marine and coastal zones			✔	✔	✔	✔	✔		✔	✔		
Atmospheric pollution			✔	✔			✔	✔		✔		✔
Other urban and industrial problems associated with contamination and waste			✔	✔	✔	✔	✔	✔		✔		✔

(a) ✔ Deteriorating environmental trend.
(b) ✔ Critically important environmental issue of concern.
Source: UNEP 1997, pp. 6–7.

- Deforestation is a major problem, due to forest clearance for commercial agriculture and the harvesting of fuel wood
- Important bio-diversity losses are occurring due to commercial exploitation of the land and its exploitation by the poor
- Water availability in Africa is highly variable. Ground water quality is deteriorating. Out of 25 countries in the world with the highest percentages of population without access to safe drinking water, 19 are in Africa
- Marine and coastal environments are significantly deteriorating due to development of tourism, over-fishing, clearing of mangrove forests, erosion and sedimentation, and rapid urbanization in coastal areas

Asia and the Pacific

- Soils suffer from varying degrees of erosion and degradation due to multiple development pressures

- Deforestation rates are very high, reflecting the pressures of high population densities and rapid population growth, land clearing and over-harvesting
- Water resources are highly variable. Extensive water pollution is caused by domestic sewage, industrial effluents and run-off from agriculture and mining
- Urban air pollution is a serious problem in many of the major cities, due to rapid growth in their energy demands and heavy dependence on coal and other low-grade solid fuels
- The region contains three of the world's eight bio-geographic realms and includes the world's highest mountain system, the second largest rain forest complex and more than half of the world's coral reefs. Both terrestrial and coastal habitat loss are major sources of concern
- The majority of the population in this region lives in coastal areas; more than one-quarter of the world's 75 largest cities are near or on the region's coast line. Of world fisheries production, 47% takes place in the region and is very vulnerable to coastal and marine pollution
- Urbanization has been accompanied by a proliferation of slums and squatter settlements without access to basic infrastructure, clean water and sanitation. Increasing affluence among a proportion of the urban population has been associated with increasing traffic congestion, air and water pollution problems, and growing quantities of industrial and hazardous wastes
- Vulnerability to natural disasters has increased due to growing urban populations and environmental degradation from man-made causes

West Asia

- More than 75% of the region is desert and an increasing proportion of the permanent pasture areas is subject to erosion because of reduced vegetation cover
- Deforestation, over-fishing, pollution, destruction of habitats and depletion of ground water all contribute to increased pressure on habitats and eco-systems
- Increased water demands are expected to lead to more extensive water shortages. Water salinity and water quality problems are of increasing concern throughout the region
- The coast and marine environments have come under increased pressure from general development, expanding tourism, increases in oil spills, and other sources
- Rapid urbanization has been associated with the loss of prime agricultural land, severe waste disposal problems and air quality difficulties

Latin America and the Caribbean

- Over 70% of the agriculturally used dry lands in South America suffer from modest to extreme degradation
- The accelerated transformation of the extensive tropical and other forests into permanent pasture and other land use is a critical environmental problem for the region
- Five of the 10 leading ecological mega-diversity countries in the world are located in this region (Brazil, Colombia, Ecuador, Mexico and Peru). These highly

valued biological reserves are threatened by a combination of habitat loss and habitat fragmentation
- Water resources are unevenly distributed within the region and a number of areas have great difficulties in meeting their water needs. Water quality problems are common throughout the region
- Of the regional population, 25% live in coastal areas (a much higher percentage in the Caribbean). At least one-quarter of the region's coastline is under threat of degradation. More than 50% of the mangroves in the region have been degraded
- Of the region's population, 75% live in cities where there are multiple environmental problems associated with air and water pollution, waste disposal and land contamination arising from economic expansion and industrialization. Mega cities, such as Mexico City, Sao Paulo and Santiago, experience particularly high levels of air pollution

Central and Eastern Europe, including the Commonwealth of Independent States

- Land degradation is extensive due to erosion, acidification and pollution
- In most areas, forest degradation is a bigger problem than deforestation and is due to acidification and destruction of forest eco-systems
- The total number of species in this region, relative to other regions, is small but the percentage of threatened species is quite large
- In a number of CIS countries, over 30% of drinking water samples failed to meet appropriate chemical or biological standards. Ground and surface water contamination has been identified as an environmental priority for action within the region
- The earlier prevalence of heavy industry, the intensive use of low quality fuels, and the reliance on older production technologies resulted in high per capita emissions and air pollution concentration levels particularly in and around urban areas. Economic recession and industrial restructuring has, in certain areas, temporarily reduced certain of these pressures. Emissions from road transport are of increasing concern
- Waste generation disposal and site contamination are significant issues in urban and industrial environments in most low and middle income countries in the region

In summary, in each of these regions, population growth, urbanization and economic development and structural change have, to varying degrees and in different ways, placed additional pressures on each of the regional environments. This has resulted in land deterioration, forest degradation, losses in bio-diversity, deterioration in water quality and shortages in water supply, deterioration in marine and coastal environments, reductions in air quality and damage to human health. The next section reviews the regulatory systems within the regions which exist to deal with these environmental pressures.

2.6   Environmental Regulatory Systems

The response of low and middle income countries to the pressures and resulting environmental conditions, described in the two previous sections, are constrained to an important degree by:

- Policy priorities – other concerns (economic, social or political) may be considered more important and be given greater attention
- Institutional capacities – the capacity of the relevant institutions to formulate, approve and implement appropriate policy responses may be limited

These considerations help to explain the incomplete nature of environmental regulatory systems in many low and middle income countries and the problems of implementation which they frequently experience. Yet, despite these difficulties, progress is being made in a number of areas, as reviewed below.

Constitutional and Broad Policy Provisions for Environmental Protection

A number of countries in Africa (e.g. Namibia, Mali, Congo), Asia (e.g. India, Vanuatu), Latin America (e.g. Mexico, Panama, Chile, Peru) and Central and Eastern Europe (e.g. Montenegro) have incorporated environmental provisions into their constitutions.

Additionally, a considerable number of countries have prepared and approved broad environmental strategies in the form of national environmental action plans, and national conservation strategies, national sustainable development strategies etc. More recently these have also been prepared for lower levels of government and public administration.

Important as these initiatives are, they have limited practical significance, beyond shaping strategic thinking, until they are translated into more specific regulatory instruments or environmental policies and programmes which are then implemented. Past experiences suggest that the commitment and capability to turn broad policies into implementable actions are often insufficient.

Establishment of Environmental Ministries and High Level Advisory Councils

Growing recognition of the importance of environmental concerns has led to a number of initiatives in capacity strengthening, through the creation of new, separate ministries responsible for the environment. In *Africa*, environmental ministries and government departments have been established in Algeria, Benin, Burkina Faso, Burundi, Cameroon, Congo, Cote d'Ivoire, Gabon, The Gambia, Guinea, Kenya, Mali, Mauritius, Niger, Rwanda, Sao Toma and Principe, Seychelles, Sierra Leone, Togo, Tunisia, Uganda, Tanzania and Zambia. An alternative approach has been to unify environmental planning and management through the establishment of semi-autonomous parastatal environmental agencies, e.g. the Environmental Council of Zambia, the Nigerian Federal Environmental Protection Agency and the Gambian National Environmental Agency.

In *Asia*, government environment ministries or general environmental director-ates have been established in several countries including China, India, Indonesia, Jordan, Lebanon, Malaysia, Oman, the Philippines, Republic of Korea, Sri Lanka, Syria, Thailand, United Arab Emirates and the Yemen. Also, in a number of countries in the *Latin American and Caribbean* region, new environmental minis-tries or ministerial secretariats have been established during the 1990s, e.g. in Argentina, Bolivia, Chile, Colombia, Jamaica and Mexico.

However, in a number of cases, these environmental ministries, agencies and secretariats are relatively new, lack sufficient status and possess insufficient num-bers of adequately trained staff. In some countries, a policy of decentralising en-vironmental regulation and administration is being pursued but its effectiveness is tempered by staff and skill shortages at regional and local levels.

Additionally, inadequate co-ordination between powerful development minis-tries and environmental ministries and their agencies continues to be a problem. In some cases, a solution is sought through the establishment of high level ad-visory councils, e.g. the Tanzania National Environment Management Council, the Gambian National Environment Management Council, the Philippines National Environmental Council and the Estonian Sustainability Development Commission. However, partly due to their advisory status, progress is very un-even. Another approach is to establish environmental sections as co-ordination units within key development ministries, e.g. in ministries dealing with agriculture and water, mineral resources and petroleum, industry and electricity, as in Saudi Arabia.

Environmental Regulations, other Policy Instruments and their Enforcement

The volume of environmental regulations and other policy instruments which have been adopted in low and middle income countries has grown substantially during the past 25 years (Wilson *et al.* 1995). Among these, the most common have been *sectoral command and control regulations* relating to different types of waste and pollution and to different environmental media. However, their effectiveness is often reduced by incomplete coverage, overlapping competencies relating to their application, ambiguities over the interpretation of their provisions, inadequacies in monitoring compliance and deficiencies in enforcement procedures and practice (see Box 2.1 for further details).

More recently, broader framework legislation has been adopted in a number of low and middle income countries. However, because of its more general character, it cannot be easily applied without more specific regulations, accompanied by addi-tional guidance on their implementation. Furthermore, if the institutions and per-sonnel involved are structured and organized on traditional lines, they may lack the capacity for its integrated application.

Some environmental standards originate in international law. Until recently, low and middle income countries played only a limited role in the formulation or implementation of international environmental agreements but they are now taking a greater interest where the financial obstacles to their involvement can be overcome.

Box 2.1   Command and Control Policies and their Implementation

Most countries in the different regions have established command-and-control policies . . ..
However, the effectiveness of implementation depends on the presence of appropriate
environmental agencies, the comprehensive scope of the legislation, availability of adequate human and financial resources to monitor compliance, and an effective legal system.
As a consequence, environmental policy implementation varies greatly between regions
and nations.

In *Asia and the Pacific* umbrella legislation and comprehensive environmental policies
increasingly provide an overall framework for regulating most forms of pollution. Specific
command-and-control policies are in place. Implementation of umbrella policies has unfortunately been hampered in a number of countries by institutional weaknesses and a lack
of human and financial resources.

Many *African* countries have also enacted umbrella legislation, mainly within the framework of national environmental action plans . . ... Unfortunately, institutional weaknesses,
lack of skilled manpower, and inadequate training facilities still hamper monitoring and
enforcement of environmental policies and regulation in a large number of African countries. Limited financial resources are partly to blame, but lack of co-ordination between
involved authorities and counter-productive government policies also contribute.

In *Latin America and the Caribbean* most national initiatives concerning the environment
revolve around command-and-control mechanisms, particularly legislation. While new
institutions established to deal with the environment greatly assist implementation of legislative reforms, institutional weaknesses remain and are often aggravated by limited
finances.

Environmental programmes, institutions and laws in *West Asia* have sometimes been
created haphazardly and not in the context of an over-arching strategic plan for the environment. This has resulted in dominantly sectoral approaches to environmental planning
without due consideration of the need for cross-cutting policies and institutional requirements . . ..

Source: UNEP 1997, pp. 139–140.·

Economic Instruments continue to attract attention, as they do in high income
countries, as potentially important instruments of environmental policy and sustainable development. However, as elsewhere, they are only used on a limited scale.

> Few, if any, economic incentives have actually replaced regulations because most have
> been introduced with the primary objective of increasing Government resources rather
> than altering behaviour towards more environmentally friendly activities. Also, some
> examples are available whereby economic instruments have been used 'for short-term
> economic gains' which resulted in degradation of the environment (UNEP 1997, p. 132)

One of the more widespread uses of economic instruments in some low and middle
income countries is in financing environmental funds. The resources of such funds
may come from the regular government budget, service fees and fines, pollution
charges and international grants. These funds may then be used to finance capital
expenditures or research expenditures relating to environmental protection. Such
funds have operated in Thailand, Burkino Faso, Chile and the Philippines, and have
been widely used in the Central and Eastern Europe region. *Environmental planning and other policy instruments* are also used to varying extents in low and middle

income countries. National environmental strategies and planning studies have been prepared in many of these countries, as have economic development studies and strategies. Local land-use planning is less well developed in most countries, although local environmental planning is undertaken for certain types of areas (urban areas, coastal zones, forests, national parks). Difficulties in plan implementation occur in many cases.

Environmental impact assessment procedures at the project level are a regulatory requirement in many countries and this is reinforced by the EIA procedural requirements of numerous development banks and bilateral aid agencies. Formal strategic environmental assessment procedures, applicable to policies, plans and programmes, are much less developed (see Chapter 3 for further details).

Newer policy instruments – environmental auditing and environmental management systems, eco-labelling and other voluntary instruments – are still in their infancy although increasing numbers of each can be found in Asia, Latin America and Central and Eastern Europe. Similar problems to those that have already been described are experienced – weaknesses in the formulation of the instruments, institutional deficiencies, insufficient trained staff, financial limitations and limited environmental commitment to the instruments concerned. Nonetheless, there is a growing consensus that certain of these instruments will have an increasing role to play in the future.

Transparency and Public Participation in Environmental Regulatory Systems

Until recently, democracy and political liberalism did not feature strongly in many low and middle income countries. These political circumstances were reflected in the limited degree of openness, access to information and public participation in their environmental regulatory systems. During the 1990s, the process of political liberalization has advanced in a number of these countries (notably in parts of Central and Eastern Europe, Sub-Saharan Africa and Latin America) and is beginning to be reflected in their environmental legislation. However, this has not been fully transposed into practice for a mixture of reasons – institutional inertia, suspicion and inexperience within the public, cultural restrictions on certain forms of public participation, low literacy and education standards, and pre-occupation with other more pressing problems.

UNEP (1997) reports that, in a number of *African* countries, more attention is being given to capacity strengthening through greater emphasis on: environmental education, training and the provision of environmental information; the role of NGOs and the media; and active participation of grassroots groups in the environmental management process (p. 144). In the *Latin America and Caribbean* region, NGOs are playing a larger role in different kinds of environmental initiatives. New laws have also been passed in Mexico, Colombia and Chile which require governments to establish environmental planning committees with broad social representation at the provincial or local levels. Environmental education programmes are being developed at all levels within the region and more environmental information is being published (p. 185).

In the *West Asia* region, the number of NGOs is growing but there is felt to be a need to strengthen their capacities and effectiveness. Environmental education programmes are also being developed in most countries in the region. However, there is a general lack of environmental data and information. In *Central and Eastern Europe*, environmental NGOs were centrally involved in the political changes which took place at the end of the 1980s. As a result, in the early 1990s, they secured some provision for access to information and public participation within the environmental regulatory system. However, with the increasing economic difficulties that followed, the participation of NGOs and the public became less (Fisher, undated) though it may be expected to increase again as economic conditions improve.

UNEP (1997) concludes:

> The past five years has witnessed the emergence of democratisation in several regions and, with it, a growth in the intertwining of politics and the environment. Many of the mechanisms required to successfully implement sustainable development rely on accountable partnerships between government, industry, business and the community at large. The value of popular participation by both individuals and interest groups in the development of environmental policy is increasingly being recognised by formerly centralised governments: thus opening the way for greater environmental citizenship in the future (UNEP 1997, p. 141)

2.7 Environmental Assessment, Environmental Regulatory Systems and Sustainable Development

EA procedures are an integral component of environmental regulatory systems in many low and middle income countries. According to UNEP, in the mid 1990s, 53 of the developing countries and transitional economies which they had studied had established some form of EIA process, either within framework legislation or in separate EIA legislation, and many other countries were working on developing such legislation (Wilson *et al.* 1996). More detailed and recent information on the extent of environmental assessment legislation in low and middle income countries is provided in the next chapter.

The importance of EA procedures, as a component of environmental regulatory systems, in promoting sustainable development is highlighted in Box 2.2. Their influence depends, however, not only on the content of those procedures but also on the effectiveness of the regulatory systems of which they form part. The interdependencies between the EA process and the environmental regulatory system in securing their mutual effectiveness are summarized below.

1. The environmental assessment process can contribute to the overall effectiveness of environmental regulatory systems in the following ways:
   a) It is the major policy instrument, within the environmental regulatory system, for integrating environmental considerations into the planning and appraisal of development activities. It is applicable at the level of development policy-making, planning and programming (in the form of strategic environmental assessment) as well as of development project planning and appraisal (in the form of environmental impact assessment).

b) It breaks out from the traditional system of single media environmental controls and substitutes an integrated, multi-media system of appraisal which is potentially more cost-effective and efficient in achieving environmental objectives.

c) By virtue of its emphasis on public accessibility to environmental documentation and its regulatory requirements relating to consultation and public participation, it should stimulate greater transparency and higher levels of community involvement and commitment in both the environmental regulatory system and the development process.

d) It can contribute to improving the environmental performance and cost effectiveness of the environment regulatory system as a whole through its use, as part of regulatory impact analysis, in the appraisal of proposed new regulations and the evaluation of existing regulatory systems (OECD 1997).

2. However, the effectiveness of the environmental assessment process depends upon the overall effectiveness of the environmental regulatory system within which it operates. In this respect, a number of the shortcomings which have been identified in current regulatory systems in low and middle income countries need to be recognized and addressed.

a) Inadequate co-ordination between environmental ministries and development ministries, which hampers the integration of environmental considerations, through SEA procedures, into the overall development process.

b) Difficulties in integrating EIA procedures into the command-and-control system for development projects because that system is, itself, not working effectively.

c) Implementation of privatization and deregulation policies with insufficient regard for their potentially damaging effects on the environmental planning and command-and-control systems to which SEA and EIA procedures are attached.

d) Institutional resistance to integrated (multi-media) forms of environmental planning and pollution control.

e) Institutional resistance to greater public access to environmental information, the transparency of environmental planning and pollution control processes and of public participation within them.

f) Deficiencies in institutional capacities, shortages of adequately trained staff and other resources, inadequate base-line data and environmental monitoring.

Box 2.2   Environmental Assessment and Sustainable Development

'The need to incorporate environmental planning into national socio-economic planning is now widely recognised. The Environmental Impact Assessment (EIA) process has become, since the 1970s, the predominant tool for such integration. Indeed, the EIA process constitutes the critical link between environment and development for it demands that the process of economic development takes into consideration the ecological perspective of socio-economic transformation . . .. The further adoption, implementation and development of EIA legislation is one of the most important emerging trends in national environmental legislation.'

Wilson P *et al.* 1995, pp. 202–205.

The main implications to be drawn from this review of the development and regulatory context of environmental assessment in low and middle income countries are the following:

- Whilst the underlying principles of environmental assessment are generally applicable, the particular form which EA regulations takes needs to be adapted to the circumstances of the particular country concerned. In certain cases, it may be necessary to adopt a step-by-step approach to EA implementation taking into account the institutional and other constraints within the regulatory systems concerned
- When introducing new EA measures, it is often appropriate to consider: (a) what accompanying changes to the environmental regulatory system would enhance EA performance; and (b) what other desirable changes to the system might be facilitated by the EA initiative?
- EA initiatives need to address potential deficiencies in practice as well as in legal enactment. Attention should be given, *in advance of enactment*, to the need for capacity-strengthening, establishing data banks and base-line monitoring systems, preparation of guidance, and awareness raising and training requirements

Discussion Questions

1. Examine how the process of economic development may place extra pressures on the environment and how the different environmental policy measures which are available might alleviate these pressures without retarding development.
2. Examine how environment assessment requirements might best be integrated into the existing environmental regulatory system in a given low or middle income country. What changes, if any, may be needed to the existing environmental regulatory system for these EA requirements to work effectively?
3. What are the main features of the development planning system and environmental regulatory system which you consider should be taken into account when formulating EA provisions for a given LDC or CIT? What are the key EA provisions you would recommend as a consequence?

Further Reading

The relationship between development and environmental quality, and between different kinds of policy instruments and their environmental, economic and social consequences, are discussed in World Bank (1992) and OECD (1997). Individual country and regional data relating to development pressure indicators can be obtained from the *World Development Report* (World Bank, annual) and, relating to environmental trends and concerns, from the *Global Environment Outlook* (UNEP 1997, biennial) and *World Resources* (World Resources Institute 1996). An overview of the status and trends in environmental regulation in developing countries is contained in Wilson *et al.* (1995) and UNEP (1997). Applications of the pressure-state-response model in individual countries in transition are to be found in the *Environmental Performance Reviews* published by OECD and UNECE (for example, OECD (1995) relating to Poland and UNECE (1996) relating to Estonia).

References

Fisher, D (undated) *Paradise Deferred: Environmental Policy-Making in Central and Eastern Europe*, Royal Institute of International Affairs, London

OECD (1995) *Environmental Performance Reviews: Poland*, OECD, Paris

OECD (1997) *Reforming Environmental Regulation in OECD Countries*, OECD, Paris

UNECE (1996) *Environmental Performance Reviews: Estonia* United Nations, Geneva

UNEP (1997) *Global Environmental Outlook*, UNEP/Oxford University Press, Oxford

Wilson, P *et al.* (1995) 'Emerging Trends in National Environmental Legislation in Developing Countries' Ch 12 in UNEP (1995) *UNEP's New Way Forward: Environmental Law and Sustainable Development*, UNEP/Environmental Law Unit, Nairobi

World Bank (1992) *World Development Report, 1992*, World Bank, Washington, DC

World Bank (1997) *World Development Report, 1997*, World Bank, Washington, DC

World Resources Institute (1996) *World Resources: A Guide to the Global Environment, 1996–7*, Oxford University Press, Oxford

3

Comparative Review of Environmental Assessment Procedures and Practice

Clive George

3.1 Introduction

As was noted in Chapters 1 and 2, low and middle income countries vary considerably in many ways, including climate, ecology, resources, political and administrative systems, social and cultural systems, and the level and nature of economic development. As well as affecting the problems which environmental assessment needs to address, these also influence the extent to which EA systems have been established in each country, their regulatory form, and their practical application.

Many countries have introduced their own EA systems, with detailed legislation that applies to all development projects that are likely to have significant impacts, irrespective of the source of funding. In other countries, either there is no EA legislation, or it remains at the enabling level. EA is then mainly carried out for activities that are financed by development banks and aid agencies, whose operating procedures require EA as a condition of funding. In some countries assessments are required under agreements with individual investors, such as oil companies. In poorer countries, externally financed projects may be the only large scale development activities being undertaken.

This chapter reviews EA procedures and practice in low and middle income countries according to geographical regions, describes the EA procedures of development banks and aid agencies, and presents an overall review and comparison

Environmental Assessment in Developing and Transitional Countries. Edited by N. Lee and C. George.
© 2000 John Wiley & Sons, Ltd.

with EA in high income countries. Tables in the Appendix to the chapter summarize the status of EA legislation in the late 1990s in those low and middle income countries for which data have been obtained (EIA Centre 1999; Donnelly *et al.* 1998), and also give more detailed information on the systems in a selection of individual countries and agencies. A large number of sources of information have been used. Rather than referencing all of these sources in the text, a principal reference for each listed country is given in the Tables. The other sources used are detailed in EIA Centre (1999). Guidance on further reading is provided at the end of the chapter.

3.2   Sub-Saharan Africa

Sub-Saharan Africa is among the least developed regions in the world, containing 60% of all low income countries. Rural development programmes, often embracing many small scale projects, play a major role in many African countries. Medium and large scale developments include road schemes, ports, water resources projects, forestry and minerals extraction. Tourism is important in several of the countries, often centred around wildlife.

South Africa is the only country with a generally applicable environmental assessment system that is well established. Although this has only recently been given the full force of detailed legislation, it has been continuously developed and applied since the 1970s and 1980s. Several other countries have introduced EIA legislation or procedures, but generally these are still at an early stage of implementation. Interestingly, in Tanzania the national parks authority (TANAPA) has developed its own EIA procedures, which apply to development activities under its own jurisdiction. The electricity supply organization has similar procedural requirements. It may be noted that in Tanzania, as in other parts of southern Africa, national parks are an important part of both the environment and the economy, and represent an acute short-term example of the inter-dependence of environment and development. In Ghana, EA procedures also have a strong sectoral emphasis, particularly in the mining industry. Elsewhere, EIA practice is mainly conditioned by the procedural requirements and capacity-building policies of development banks and aid agencies.

Screening, Scoping and Assessment

With the exception of certain countries such as South Africa, screening tends to be based largely on the requirements of funding agencies, or licensing agreements with international companies. In South Africa a mandatory scoping report must be prepared for most of the projects included in a screening list. Elsewhere, even where national legislation does exist, it is often applied mainly to internationally funded projects. Nigeria, for example, has a detailed EIA system enacted by Presidential decree, with discretionary provisions relating to its implementation. These have mainly been applied to projects for which EIA is required by donor agencies or international oil companies.

In many countries in the region the scope of the assessment has to be approved by the competent authority, although it is normally drawn up by the consultants carrying out the assessment. In Tanzania, TANAPA defines the scope itself, and appoints consultants to undertake the assessment. The assessment is paid for by the developer, and so the system places a strict financial limit on the cost. Because many parks projects are relatively small, the cost is set fairly low, and this has sometimes made it difficult to assess impacts in sufficient depth. However, a similar need to tailor the EIA system to the specific development needs of the country has also been identified elsewhere in the region. In the Seychelles, this has been expressed in relation to the particular needs of small island states.

In Angola and Burkina Faso the assessment is carried out in part by government officials, while in Uganda government approval is required for the choice of EIA consultants. Elsewhere the selection of consultants is generally at the developer's discretion. In some countries such as South Africa, Tanzania and Sudan local consultants are widely used, but in the region as a whole there is considerable reliance on foreign consultancies.

Public Involvement

In South Africa public participation is mandatory in the preparation of the scoping report and in carrying out the full study. The public consultations which have been undertaken must be fully described in a scoping report and in the final EIA report. The EIA report becomes a public document, but not until after the approval decision on the environmental acceptability of the project has been made. These provisions reflect a general move towards participatory democracy, from a more authoritarian tradition of decision-making. Elsewhere in the region, public involvement is often only undertaken to meet the requirements of funding agencies.

Decision-making Structures

In Tanzania the economic sectoral authorities are responsible for deciding the environmental acceptability of the project, and in Ghana sectoral guidelines are applied by the relevant ministries. In South Africa the responsibility rests with the national environmental authority, but except for projects of national significance it is generally delegated to provincial governments. In most other countries in the region the competent authority for deciding or advising on the environmental acceptability of all projects is a national environmental authority. In several countries the final decision on project approval is taken by a cross-sectoral committee or review body.

Strategic Environmental Assessment

Several ad hoc strategic environmental assessments have been carried out in South Africa, including a regional SEA within KwaZulu-Natal. In Tanzania strategic assessments are part of the planning process for national parks. Elsewhere in the region SEA is not well developed.

3.3   East Asia and the Pacific

One of the most notable characteristics of East Asia is its rapid industrialization and, despite a recent slow-down, its general economic growth. Major activities, as well as industrial developments, include extensive development of forests, coastal developments, and large scale water resources and other infrastructure projects. These pose particular challenges for EA, in handling complex social issues and potential loss of biological diversity. China's Three Gorges dam entails the resettlement of between one and three million people. Water resources projects in the Mekong river basin and elsewhere introduce the additional complexity of trans-boundary impacts. Efforts to deal with the global issue of biodiversity conservation include a policy level environmental assessment for forestry management in the Philippines.

The more rapidly developing countries (Malaysia, South Korea, Taiwan, China, Indonesia, the Philippines and Thailand) all have their own environmental assessment systems. Other countries in the region have developed less rapidly, and have not experienced the same degree of industrialization or economic growth. These include Cambodia, Laos, Myanmar and Vietnam. Vietnam introduced its own legislation in 1994 but, in general, EIA in these other countries is influenced strongly by development funding agencies. Papua New Guinea has introduced enabling legislation, but comprehensive EA systems have yet to be built up in any of the Pacific island states.

Screening, Scoping and Assessment

Some countries such as Taiwan have developed screening and scoping systems which take into account the specific characteristics of individual projects. Such systems may require a high level of expertise from administrative staff. Some of the less developed countries in the region rely more heavily on standard screening lists of project types, and scoping lists of the types of impact that must be considered in assessments. Malaysia has developed a comprehensive set of sector-specific guidelines, and uses a system of preliminary appraisal as an extension to this.

Many of the countries in the region including China, Malaysia, the Philippines, Taiwan and Thailand have some form of state licensing or accreditation of consultants who undertake EIAs.

Public Involvement

Most of the countries' EIA systems contain formal provisions for some degree of public involvement in the EIA process, although in some cases this tends to be rather general and limited in its details. In a number of countries EIA reports do not need to be made readily available to the public, and public participation in the EIA process as a whole is relatively limited.

Decision-making

In several countries (China, South Korea, the Philippines and Taiwan) EIA systems are fairly highly decentralized, with authority being delegated to local government or provincial environment agencies. In China this partly reflects the sheer geographical size of the country, but also the structure of the country's development planning mechanisms. In Malaysia development decisions for many types of project are now taken by state (i.e. regional) government administrations, but the original EIA system was administered entirely at the federal level. Between 1993 and 1995 the system was gradually decentralized, partly to reduce the administrative load at the Federal level, and also to allow better integration with local development planning processes. A similar process of decentralization is under way in Indonesia.

Strategic Environmental Assessment

Formal provisions in the region for strategic environmental assessment are limited. However, SEA has been applied in China, Korea and Taiwan for regional development plans, and also to certain policies. The Asian Development Bank is preparing a strategic environmental framework for the Greater Mekong, through its Office of Environmental and Social Development in the Philippines.

3.4   South Asia

The South Asia region has not experienced the same rapid industrialization as East Asia, although in India industrial development has been considerable. Overall, the region has the lowest average per capita income, with high rural and urban population densities. There are wide variations in both geography and climate, from the Himalayas to low lying tropical coastal regions and islands. In monsoon areas there are large seasonal climatic variations.

Water resources are important in the region. Both large scale and small scale projects have attracted controversy. Forestry is also important in several of the countries, and is associated with long standing traditions of environmental concern. Industrial development and energy projects are common in some areas, particularly in India, while coastal development is significant in most countries in the region.

Except in the case of India, environmental assessment procedures and practice are strongly influenced by development banks and aid agencies. India, which is the largest country in the region, is the only one with a long established EIA system. Sri Lanka and Nepal have implemented detailed regulations more recently, but experience with them is still relatively limited. Pakistan's EIA system was introduced by Presidential decree in 1983, but to date it does not have full legal force. Bangladesh and Bhutan have introduced enabling legislation relatively recently.

Screening, Scoping and Assessment

The Indian screening system is based on lists of projects, with thresholds, for which EIA is mandatory. Some of the thresholds are expressed in terms of project cost. A distinction is drawn between a 'rapid' EIA and a 'full' EIA, which requires a 12 month baseline assessment, to take account of what may be considerable seasonal variations in climate, flora and fauna, and agricultural activities. The Nepalese screening system allows for an Initial Environmental Examination, which may lead to a full EIA.

Scoping in India is based on sector-specific guidelines, and is less influenced by the individual characteristics of a project and its environment. This, alongside the provisions for a full EIA, can result in EIA reports with very extensive baseline descriptions. The more recently introduced Sri Lankan system includes guidelines for the scoping process, and provisions for scoping meetings.

In most of the region, EIAs are undertaken by consultancy organizations, who do not need to be licensed for this purpose.

Public Involvement

Until recently, public participation played a relatively minor role in appraisal and decision-making in the Indian EIA system. However, in 1997 public hearings were made mandatory. They are organized by the relevant State authorities, before the EIA report is submitted to the central government competent authority. Under the Nepalese regulations, the EIA report must be made accessible for public review.

Decision-making

India's EIA system is relatively centralized, despite the country's size. It is administered primarily at central government level, with review of EIAs being undertaken by national panels of technical experts. In 1997 a greater degree of responsibility for environmental decision-making for some project types was delegated to individual State Pollution Control Boards.

In Sri Lanka the EIA regulations enable the Central Environmental Authority to delegate its powers for administering and approving EIAs to the sectoral ministries responsible for development planning. This can enable a close integration of environmental assessment into the planning process, but has led to some difficulties, through limited environmental expertise and financial resources in the relevant ministries.

Strategic Environmental Assessment

Formal provisions for strategic environmental assessment are not well developed in the region. Ad hoc assessments include an SEA carried out in Nepal in 1995 by IUCN, for the Bara Forest Management Plan. In India strategic environmental issues are being addressed by the Central and State Pollution Control Boards, through the preparation of coastal zone management plans and zoning plans for industrial development.

3.5   Central and Eastern Europe and Central Asia

Although most of the transitional economies in this region fall into the low or lower-middle income groups, a number of them are fairly highly industrialized.

Pollution from old and inefficient industries has been a major problem. In some countries economic difficulties have caused industry to contract, and pollution has fallen as a result, but the revitalization and environmental improvement of industry remains an important development priority. Some of the remoter areas have unique ecological characteristics important for biodiversity conservation.

Minerals extraction is a major activity in many areas, including some of the less highly industrialized countries of Central Asia. Other major development activities include forestry, water resources projects and a full range of infrastructure developments.

The region's previous political, economic and planning systems have had a major influence on the development of EA. The current transition to more market-based economies and more open democratic systems has led to changes in EIA systems, to varying degrees in different parts of the region.

Environmental assessment became widely adopted as a regulatory requirement in the region during the period of Soviet influence. However, the form of assessment that was developed, particularly in the countries of the former Soviet Union, differed from that generally adopted elsewhere in the world. In these countries, procedures for State Environmental Review (SER, alternatively referred to as State Ecological Expertise) were introduced, placing the prime responsibility for assessing the potential impacts of a proposed development on state environmental authorities, and committees established by them. For their part, developers (usually state enterprises or other state organizations) were often required to include an assessment of environmental impacts (OVOS) in the project documentation they submitted for SER. Further details of this SER/OVOS approach are given in Chapter 12 (Section 3). Similar legislation was also introduced in other countries of the region, although in some cases (e.g. Poland) containing features adopted in the west.

The SER/OVOS approach has been retained, with adaptations, in many countries of the former Soviet Union, particularly those which have become members of the Commonwealth of Independent States (CIS). However, following the dissolution of the Soviet Union and the move towards more market-based economies, many other countries in the region have introduced systems more akin to those of Western Europe. In a number of cases, their goal is to achieve closer approximation to the EA requirements of the European Union.

Screening, Scoping and Assessment

In general, SER applies to any development activity requiring planning approval. In this sense there is no screening process, but since the competent authority itself carries out a major part of the assessment, it is in a position to decide on a case by case basis how extensive the assessment needs to be. For major projects, or any likely to have particularly significant impacts, an in depth analysis may be conducted, and an OVOS submission may be required according to a screening list. For others, the review may be no more than a check against development policies, pollution standards and other norms, similar to those carried out in other countries for simple planning approvals and pollution consents. The system is oriented strongly towards pollution control, and scoping is based largely on legal standards and norms.

Under the EA systems introduced in other (non-CIS) countries of the region, responsibility for conducting or commissioning the assessment usually rests with the project proponent, as in most western countries. More clearly defined screening procedures are therefore necessary. In Estonia the EIA is organized by the competent authority, but at the proponent's expense, and so again a clear screening procedure has been defined. Proposed Estonian legislation, currently in draft form, passes the responsibility for organizing the EIA to the developer. Under these systems, assessments are normally carried out by professional institutes or consultancies. A requirement for state recognition of expertise has been retained in many countries, through systems for state licensing of individuals or organizations approved to conduct EIA studies. A variety of scoping methods are used, including general and specific checklists, preliminary assessment and approval by the competent authority

Public Involvement

Throughout the region, environmental NGOs have been in the forefront of promoting the whole process of democratization. As well as being incorporated into recent EA systems, certain provisions for public involvement have been included in the SER process. Several countries' SER systems allow for a Public Environmental Review (PER) to be organized, involving public hearings, particularly when there is public or NGO pressure. The findings of the public review are considered by the state environmental review. However, neither the competent authority's full assessment report, nor the developer's OVOS submission, are normally made readily available for public comment.

In many of the non-CIS countries that have developed new EA legislation, broad public participation requirements have been introduced that are similar to those of high income market economies. Actual practice is still evolving however, and in some cases, public access to EIA reports is not easily obtained.

Decision-making

In many countries of the region, particularly the larger ones, the EA system is fairly highly decentralized and integrated with local government planning systems.

Strategic Environmental Assessment

Under the region's former central planning systems, a less clear distinction was drawn between projects and plans, since government was responsible for both. As a result, several of the countries have provisions in their legislation for a form of strategic environmental assessment as well as for project level assessment, which mainly relates to development plans. In practice, the EA of policies, plans and programmes is much less developed than for projects, as in the rest of the world. Some countries, however, such as Slovakia, have developed guidelines and are building up significant experience.

3.6   Middle East and North Africa

Many of the Middle Eastern and North African countries share a broadly similar climate, a common language, and a cultural and religious heritage that is characteristic of the region. The population in most of the region is concentrated in the fertile coastal areas and river valleys, where both agricultural activity and several large and rapidly growing cities are located. High population densities in some of these areas exacerbate the problems of air pollution and water pollution associated with increasing urbanization and industrialization. Water availability places a significant constraint on development. One of the biggest determinants of national economic prosperity is oil. Some oil states (as well as Israel) are high income economies, while others are classified as upper-middle income. Yemen is the only low income economy, and the majority of countries in the area are in the lower-middle income category.

Water resources development, coastal development (including tourism) and industrial development are significant activities in many parts of the region.

Among the developing countries in the region, only Oman, Tunisia and Turkey have long established EA systems. Egypt and Morocco have enacted EA legislation more recently. Tunisia's economy, like Turkey's, is more diverse than many in the region, with industrial growth running at around 5% per annum. Enabling legislation for EIA was introduced in Tunisia in 1988, followed by detailed provisions in 1991. In Oman and Turkey enabling legislation was introduced in 1982 and 1983, respectively, and detailed provisions in 1993. Detailed legislation was introduced in Egypt in 1994, but was not fully implemented until 1998.

Legislation in other countries is still mainly at the enabling stage or in draft form, and EA practice is often conditioned by overseas agencies' requirements.

Screening, Scoping and Assessment

Tunisia, Turkey and Egypt have screening lists allowing for different forms of preliminary assessment, as well as for full mandatory EIAs. In Tunisia and Turkey, the scope of a full assessment is defined by the competent authority, using standardized formats. Egypt has sectoral guidelines for full EIAs, while for 'grey list' projects a limited scope is defined by the competent authority.

Egypt has a system of licensing consultants to review EIA reports for the competent authority, but does not require a licence for conducting an assessment.

The EIA system in Oman differs from that in most other countries outside the CIS countries of the former Soviet Union, in that responsibility for the assessment is placed primarily on government officials. For large projects, the competent environmental authority may (at its discretion) require the developer to submit a full EIA report. Otherwise the authority evaluates the project itself, on the basis of a form completed by the proponent. This details, in particular, materials used and pollutants discharged. The authority may undertake a site visit as part of the evaluation.

Public Involvement

Turkey's EA system includes provisions for public hearings. Elsewhere in the region, EIA is widely perceived as a technical process exclusively involving professional experts. There is not a strong tradition of public participation, which tends to take place primarily in cases of public controversy, or in response to the requirements of overseas funding agencies. Egypt's official guidelines require the views of NGOs and affected groups to be obtained as part of the EIA. However, as in many other countries in the region, the EIA report is generally regarded as a confidential document, unless a funding agency is involved and requires it to be in the public domain.

Decision-making

The Tunisian EA system is centralized, with little devolution of powers below central government. The country is, however, a relatively small one (nine million population). Turkey and Egypt are much larger (over 60 million people), but their systems are also centralized. In Egypt, a high degree of consultation takes place during the approval process, between the national environmental agency and the appropriate planning body. Depending on project size and nature, this is either the relevant sectoral ministry or the appropriate Governorate (local government). The Tourist Development Authority, in particular, has developed extensive guidance, and contributes strongly to the implementation of EIA, as a key component of its strategy for the sustainable development of the industry.

In Jordan, detailed procedures have been prepared by the Aqaba Region Authority, in association with the World Bank, but may be superseded by detailed national legislation currently in preparation. In Lebanon, procedures are being developed by sectoral development ministries. Elsewhere in the region, decision-making is relatively centralized under the control of a national environmental authority.

Strategic Environmental Assessment

A form of strategic environmental assessment has been applied to coastal development in Egypt. In general, experience with SEA in the region is limited, and it is not required by legislation.

3.7   Latin America and the Caribbean

Parts of Central and South America, and also some of the Caribbean states, have experienced rapid industrialization similar to that of East Asia. These include the two largest countries in the region, Brazil and Mexico, both of whose economies fall in the upper-middle income group. This group also includes Argentina, Chile, Uruguay and several of the Caribbean states. The majority of other countries in the region are lower-middle income economies. Forestry development, coastal development and major water resources projects are common in the region. Oil and other mineral extraction are important in some of the countries.

The pattern of development of EA in the region parallels that of East Asia, with the larger, more industrialized, and more economically advanced countries (notably

Brazil and Mexico) having more fully developed EA systems. Venezuela has also introduced detailed legislation. A national code for EA was introduced in Peru in 1990, under which sectoral development ministries have issued their own regulations. In Chile detailed regulations have been introduced more recently, but many assessments were carried out under earlier administrative provisions. Assessments have also been carried out voluntarily in Argentina, in addition to mandatory EAs for major hydroelectric schemes. Among the poorer countries Belize, Bolivia and Costa Rica have developed EIA regulations. In several other countries in the region legislation remains at the enabling level, and the development funding agencies have a stronger influence on practice.

Screening, Scoping and Assessment

Most of the countries' systems use screening lists. In Brazil and Mexico, these define which types of project are subject to EIA at the federal level, and which types are handled at the state level. In Venezuela screening also takes account of whether the project is in an environmentally sensitive area. Several countries define a general scoping checklist in their legislation, while procedures in both Brazil and Belize give scoping guidance. In Belize the assessment is carried out by government officials, but elsewhere it is generally conducted by consultants. Mexico and Peru both have systems for certification of EIA consultants.

Public Involvement

Provisions for public participation tend to be relatively stronger in the higher income countries in the region. Under Brazilian federal law EIA reports must be made available for public access, and public hearings take place for controversial projects. Belize and Peru also have provisions for public meetings. In Mexico the public has rights of access to information, but is not normally able to consult the files until after the decision has been made. In Peru and Chile the EIA report is available to the public. Public involvement in earlier stages of the process tends to be discretionary in most countries in the region.

Decision-making

A degree of decentralization of authority for deciding on the environmental acceptability of a project is common in the region.

In Brazil and Mexico, federal laws and regulations are supported by legislation in individual states. The EIA systems in both countries are fairly highly decentralized for projects not of national importance, allowing close integration with local government planning systems. In Brazil around 40% of all EIAs are in Sao Paulo, which is the most populous of the states, and highly industrialized. In Mexico some difficulties were initially experienced in developing sufficient expertise in all states to operate a decentralized system. However, this situation is improving with increasing experience, and state competence has been further assisted by recent developments in computer-based training and management systems. In Chile, regional Commissions for the Environment have been established, as well as a national Commission.

In Ecuador, Uruguay and Venezuela, responsibility for defining and administering detailed EA requirements rests with sectoral development ministries. In Peru, detailed regulations have been issued by sectoral ministries and also by the Municipality of Lima.

Strategic Environmental Assessment

Strategic environmental assessment has been introduced formally under the Sao Paulo legislation in Brazil, which requires the assessment of certain development plans. Local and regional urban zoning plans are subjected to environmental assessment under the Chilean system. Elsewhere in the region formal provisions for SEA are less well developed.

3.8   Development Banks and Aid Agencies

The principal development banks are the World Bank (which operates globally), the African Development Bank, the Asian Development Bank, the Caribbean Development Bank, the European Bank for Reconstruction and Development, the Inter-American Development Bank, the Islamic Development Bank and the Development Bank of Southern Africa (which operate in their respective regions). Much of these banks' funding comes from high income economies with global or regional interests, as well as countries in their own region. The European Investment Bank is funded by Member States of the European Union, and operates primarily in those countries where member states have historical or other interests. The banks provide loans for development activities, at preferential interest rates. They also give grants, in some cases for environmental assessments, and for related capacity-building activities.

Many high income countries have their own overseas development aid agencies, which donate funds for development activities. In addition, the Member States of the European Union provide similar funding jointly, through several Directorates General of the European Commission.

During the 1970s and 1980s, several major projects supported by these banks and aid agencies attracted considerable criticism on environmental grounds, including public criticism within the countries which provide the funds. As a result, many have developed their own environmental assessment procedures (see examples summarized in Table 3.8). In some cases these are required by national legislation in the donor country (Australia, Canada, Italy, the Netherlands, USA), while in others they are based on government policy or the agency's own policies. A broad overview of these is provided below, and further details are contained in Chapter 14.

Capacity building for EA is also promoted by many other international organizations, including several UN bodies (notably the United Nations Environment Programme, the United Nations Development Programme, and the United Nations Economic Commission for Europe). The World Conservation Union (IUCN) promotes EA capacity strengthening throughout South Asia and the rest of the Asian region, from an EIA Centre in Nepal.

Relationship to National EA Systems

As part of their capacity building policies, the banks and aid agencies generally prefer recipient countries to take responsibility for EAs of funded projects, ideally using national procedures that are compatible with their own. A single assessment report can then be presented for approval by both the national authorities and their own staff. In countries where EA is new, the agencies have often supported the establishment of appropriate national procedures and administrative capacity. In such countries, the requirements of national procedures tend to have similar features to those of the agencies.

In some other countries, notably the CIS countries of the former Soviet Union, integration may be less readily achieved. The developer's OVOS submission may be too restricted in scope to meet funding agency requirements, and the SER analysis and conclusions are not normally reported in a suitable form. Conversely, assessment reports produced under agency procedures tend not to meet the requirements of national legislation and procedures. As a result, two separate assessments are often conducted in these countries, and in others whose EA procedures are distinctly different from those of the agencies.

Difficulties can also occur when more than one agency supports a project, and EA consultants have to ensure that their studies meet the detailed requirements of each agency involved. The OECD's Development Assistance Committee has addressed this problem by providing a summary of and guidance on bilateral agencies' requirements (see Chapter 14, Section 3).

Screening, Scoping and Assessment

Most agencies have screening procedures which are similar in principle to those of the World Bank, described in Chapter 14, Section 1. They differ from those of most national systems, in that they allow a greater degree of discretion to bank or agency staff. Screening lists tend to be for guidance only. They generally indicate typical projects or other activities for which a full assessment is likely to be needed, others for which a simpler environmental appraisal may be adequate, and those not likely to require either.

Responsibility for arranging the assessment normally rests with the recipient organization (usually a development ministry or other government authority). However, the assessment is often paid for by the bank or agency. Hence, Terms of Reference for the work that will be done are needed, and must be agreed by the agency (see Chapter 14). Before these Terms of Reference can be drawn up, the overall scope of the assessment needs to be defined and approved. However, the Terms of Reference can include provision for more detailed scoping (and public consultation) within the assessment itself.

To ease the task of agency staff and/or the recipient organization's staff in preparing Terms of Reference, many agencies have prepared extensive sectoral scoping guidelines. These are supported by the results of a review of individual project characteristics and of the receiving environment. This normally involves agency staff and should include site visits, at least for major projects.

Although the recipient organization is normally responsible for arranging the assessment, the agency normally also approves the choice of consultants. Agencies' policies generally encourage the use of local consultants, which can be particularly valuable in building local expertise and experience. However, for the overall management of the assessment project or for particular technical expertise, international consultants are often employed. When national systems require national certification of consultants undertaking assessments, this may entail foreign consultants obtaining national certification, or arranging an appropriate distribution of responsibilities between foreign consultants and their local partners.

Public Involvement

Most agencies' procedures encourage public involvement during the assessment itself, though the detailed form that this should take is often not specified. The procedures of the Netherlands aid agency include formal requirements for public participation. For community development projects and those involving indigenous people or involuntary resettlement, the World Bank has specific requirements and guidelines. Most agencies encourage or require publication of the assessment report, with opportunity for comment before the agency makes its decision. However, a number of agencies' procedures include the proviso that the report will not be made public without the recipient organization's permission.

Decision-making

Most development banks and aid agencies have an appraisal and decision-making structure which allows residual environmental impacts to be considered alongside social and economic factors. Typically, the agency has an environment department responsible for overseeing the EA process and ensuring that an adequate EIA report is submitted, and that appropriate design and mitigation features have been included in the project. The environmental assessment, the economic appraisal, and any other assessment (such as a separate social impact assessment or health impact assessment) are then reviewed by the decision-making body, so that all these aspects can be weighed against each other in deciding whether or not to commit funds to the proposal. If the project is for the modification of an existing operation (particularly industrial ones), an environmental audit may also be required before funding is approved, in order to identify any existing environmental liabilities (see Chapter 11).

Strategic Environmental Assessment

Increasingly, agencies' environmental assessment procedures, particularly in the World Bank, are being applied to plans, programmes and loans of a strategic nature, as well as to funded projects. The World Bank has issued guidance on EA of sectoral loans and regional development loans, and several of these have now been carried out (see Chapter 14, Section 1). The guidance includes mechanisms for tiered assessment, such that individual projects carried out under the loan agreement may be subject either to a full individual EIA, or to a simpler form of environmental appraisal guided by the overall EA for the loan agreement.

3.9   Comparative Overview

While environmental assessment has tended to evolve in different ways in different parts of the world, a number of general conclusions may be drawn.

Adoption of EA

In many low income countries, particularly the smaller ones, environmental assessment tends to be undertaken primarily where this is a requirement of development funding agencies. Increasingly, however, the assessment of such projects is carried out under national procedures and administrative systems, as well as those of the funding agencies. This is largely as a result of the policies adopted by the agencies, and the capacity-building programmes which they support. Certain funding agencies also provide support for a strategic environmental assessment of regional or sectoral programmes.

Indigenous EA systems, which apply whatever the source of a project's funds, are most common in middle income countries, especially upper-middle income ones, and where industrial development is significant (e.g. in parts of South East Asia and Latin America).

Some low income countries have also developed indigenous EA systems which operate independently of external funding. These include:

- Certain larger countries (e.g. China and India)
- Countries where adverse environmental impacts have already impeded development (e.g. Tanzania)
- Countries with a tradition of central planning (e.g. Armenia and Vietnam)

Screening

In countries with the least developed EA systems, screening may be at the discretion of senior government officials, high level committees, individual ministers or the head of state (e.g. in several countries of Sub Saharan Africa). In countries with the most developed systems, a degree of flexibility in screening is achieved by allowing some discretion to less senior officials or technical committees (e.g. Taiwan). In between these two extremes, screening systems tend to be based on lists and leave less room for discretion and flexibility.

In CIS countries, where central planning systems were previously most firmly in place, EA in the form of State Ecological Expertise is generally mandatory for all development activities which require planning approval. In such cases, there is no formalized and structured screening system for SEE, though there may be one for supporting OVOS submissions, based on the use of lists.

Most other countries with their own EA systems provide screening lists of projects for which EA is mandatory, with clear thresholds and/or criteria. A second list of projects (such as in the European Union's Annex II), for which screening may be at the discretion of local officials, is less common than in high income countries. However, many countries' use an alternative approach, of providing a list of projects for which a preliminary assessment must be carried out, to provide information for pollution control, and determine whether a full assessment is needed.

Scoping

In many low and middle income countries EA is strongly oriented towards meeting pollution control requirements. Consequently, pollution effects may be given the major consideration in scoping. In some countries the scope of the assessment must be approved by a national competent authority (e.g. Turkey and Tunisia). This is similar to funding agency systems, and in some high income countries. Where the number of assessments is larger and authority for scoping has to be delegated, sectoral checklists or guidelines are often issued, to help in defining the scope of the assessment for various project types (e.g. Malaysia). In some countries (e.g. Tunisia) standardized Terms of Reference are used for all projects of a given type, with limited requirements for individual scoping of individual projects.

Assessment

Responsibility for carrying out the assessment may be placed on:

- The developer (and consultants chosen freely by the developer)
- Consultants or other experts licensed by the state
- State experts

In most low and middle income countries the assessment is carried out by consultants employed by the developer, as in most high income countries. State licensing of consultants is more common in East Asia and the former socialist countries of Central and Eastern Europe than elsewhere.

In CIS countries and a small number of others, a partial assessment may have to be submitted by the developer along with other project documentation, but the overall assessment is carried out by the competent authority's own experts. In general, carrying out an environmental assessment is seen as the responsibility of the project proponent, which in centrally planned economies was the state itself.

Public Participation and Openness of the Process

Although many countries' EA systems make some reference to public participation, the requirements are often very general and lack detail, with resulting difficulties in their practical implementation.

Some countries make specific legal provisions for public access to EA reports, and for public comments and/or public hearings. However, many do not. Also, even when provisions are made for public hearings, the full details of the environmental assessment report are often regarded as commercially confidential, or as state secrets. In countries where the private sector plays a larger role in the economy, with less state supervision, there is a tendency towards less sensitivity to public criticism of development proposals, and greater openness in decision-making.

Consultation, Review and Decision-making

Arrangements for consultation with other government bodies, for review of EIA reports and for decision-making vary widely, according to the overall administrative structures of the country.

Many developing countries have powerful sectoral development ministries. In some cases these ministries act as the competent authority for EA (e.g. Peru and Tanzania), but more generally EA is the responsibility of a separate environmental ministry or agency whose purpose is to provide an environmental check on the activities of development ministries. Commonly, the sectoral ministry is responsible for the final approval of a project, and approval of the EIA by the environmental agency is a precondition. In other cases, final approval is by a cross-sectoral review committee or commission, which weighs environmental against other factors. In some countries (e.g. Egypt) the competent authority for EA is the country's sole environmental agency. Approval of the environmental assessment may then serve as a composite environmental permit, covering pollution control as well as other potential impacts.

In some countries (e.g. Poland) elements of development planning and/or land-use planning are delegated to local government (particularly for activities that are not of national importance). Where this is the case the decision-making structure for EA may also be decentralized, either through local offices of the national environmental agency, or through an environmental department of local government. In some countries, however, the decision-making body for EA remains at national level, even for projects subject to local planning decisions.

Monitoring

Requirements for some form of monitoring plan or environmental management plan, to be submitted with the EIA report or as part of it, are reasonably common in low and middle income countries. There may also be provisions for follow-up monitoring by the competent authority. In some cases these requirements reflect the influence of funding agencies such as the World Bank, whose own procedures are similar. In others, with a single agency responsible for all environmental matters, it reflects a degree of overlap between EA, pollution control and other regulatory activities. However, where requirements for follow-up are defined, they often lack sufficient detail to be fully effective (see Chapter 11).

Strategic Environmental Assessment

Formal provisions for some form of environmental assessment of policies, plans or programmes are fairly common in the transitional countries of central and eastern Europe. Under the former central planning systems in certain of these countries, the planning of individual projects was part of the overall planning process, so that broadly similar environmental measures were applicable to both. However, implementation was limited by a lack of clear screening processes and assessment methodologies. Although in some of these countries SEA is as advanced as in many high income countries, and more so than in some, actual practice is still evolving. Elsewhere, a few countries have formally introduced various types of strategic assessment but, in general, SEA tends to be undertaken on an ad hoc basis. As in high income countries, procedure and practice for SEA are still evolving.

Future Development of EA

The factors underlying both the adoption of EA and differences in approach and implementation are fairly complex, and vary considerably from region to region and country to country.

Many of these factors are closely related to the development process itself. When environmental legislation is relatively new, priorities may lie with immediate problems, such as the control of pollution or basic landuse planning. Limited public funds often restrict the administrative capacity of environmental agencies. A lack of development often implies the lack of a well developed market economy, with greater reliance on state planning of projects subject to EA, and consequent differences in decision-making mechanisms. Democratic processes are often not highly developed, which inhibits public participation in decision-making. Technical expertise may be concentrated in academic and technical institutions, rather than in consultancy organizations adapted to conducting EA studies.

Training and institutional strengthening programmes, such as those sponsored by development banks and aid agencies, can make a major contribution to building effective EA systems in low and middle income countries. They must, however, take full account of the infrastructure within which the EA system will operate. Academic institutions, professional institutions, non-governmental organizations, government bureaucracies, and political and economic processes, are all part of the infrastructure which contributes to EA and its effectiveness. Their own institutional development, and hence the development of EA, is partly dependent on the success of the country's overall socio-economic development.

To be effective in promoting sustainable development, the EA system must be compatible with the evolving infrastructure which surrounds it, with its capacity to support it, and with a country's own specific environmental and developmental needs. With this proviso, however, the basic principles of environmental assessment are the same everywhere. In many low and middle income countries there is much to be done before these principles can be implemented in full, but the closer they are adhered to, the more effective the development process itself is likely to be.

Discussion Questions

1. In which parts of the world is environmental assessment in low and middle income countries most/least strongly established? How can the differences be explained?
2. How do the features of environmental assessment systems vary between different regions and different countries? How might these variations be explained?
3. What barriers exist to the introduction and operation of effective indigenous EA systems in low and middle income countries? What are the respective roles of: (a) the countries' governments; and (b) international banks and bilateral aid agencies in surmounting these barriers?
4. What factors influence the degree of public participation in the environmental assessment process in low and middle income countries? To what extent, if any, does lack of public participation limit the value of environmental assessment as a tool for sustainable development?

Further Reading

A directory of impact assessment guidelines in countries throughout the world is provided in Donnelly *et al.* (1998). A comparison of EA in industrial and developing countries is given in Goodland and Edmundson (1994), together with reports on specific issues in various regions and countries. United Nations Environment Programme (1996) discusses trends and current practice in environmental assessment in developing countries, with an overview of the key principles involved. A review of environmental assessment legislation in developing countries is given in UNEP (1995), and the full text of EIA legislation in African countries is to be found in UNEP/UNDP (1996–1998). Environmental assessment in African countries is reviewed in Goodland *et al.* (1995). Bellinger *et al.* (1999) provide details of environmental assessment systems in 19 countries of Central and Eastern Europe, with an overview and a comparison of their principal features. A review of EIA in Asia is given by Lohani *et al.* (1997), together with numerous case studies. The EA procedures of bilateral aid agencies are summarized in OECD (1996).

APPENDIX   Details of EA Systems in Low and Middle Income Countries

Table 3.1   *EA Legislation in Low and Middle Income Countries*

Sub-Saharan Africa		Malaysia	d	Slovenia	d
Angola	e	Myanmar	n	Turkmenistan	d
Botswana	n	Palau	d	Ukraine	d
Burkina Faso	e	Papua New Guinea	e	Uzbekistan	d
Burundi	n	Philippines	d		
Cameroon	e	Solomon Islands	n	*Middle East and North*	
Cape Verde	e	Taiwan	d	*Africa*	
Comoros	e	Thailand	d	Algeria	e
Congo	e	Tonga	n	Bahrain	e
Cote d'Ivoire	e	Vanuatu	n	Egypt	d
Eritrea	n	Vietnam	d	Jordan	e
Ethiopia	n	Western Samoa	n	Lebanon	n
Gabon	e			Morocco	e
Gambia	e	*South Asia*		Oman	d
Ghana	d	Afghanistan	n	Palestinian Authority	n
Guinea	e	Bangladesh	e	Saudi Arabia	n
Kenya	n	Bhutan	e	Syria	n
Lesotho	n	India	d	Tunisia	d
Liberia	n	Maldives	n	Turkey	d
Malawi	e	Nepal	d	Yemen	e
Mali	n	Pakistan	e		
Mauritius	e	Sri Lanka	d	*Latin America and the*	
Mozambique	e			*Caribbean*	
Namibia	e	*Central and Eastern*		Argentina	d
Niger	e	*Europe and Central Asia*		Bahamas	n
Nigeria	d	Albania	e	Barbados	n
Senegal	d	Armenia	d	Belize	d
Seychelles	e	Belarus	d	Bolivia	d
Sierra Leone	n	Bulgaria	d	Brazil	d
Somalia	n	Croatia	d	Chile	d
South Africa	d	Czech Republic	d	Colombia	e
Sudan	n	Estonia	d	Costa Rica	d
Swaziland	d	Georgia	d	Ecuador	e
Tanzania	n	Hungary	d	Haiti	n
Uganda	e	Kazakhstan	d	Honduras	n
Zambia	d	Kyrgyzstan	d	Jamaica	e
Zimbabwe	e	Latvia	d	Mexico	d
		Lithuania	d	Paraguay	d
East Asia and Pacific		Macedonia	n	Peru	e
Cambodia	n	Moldova	d	St Kitts and Nevis	n
China	d	Mongolia	d	St Vincent and the	
Cook Islands	n	Montenegro	d	Grenadines	n
Fiji	n	Poland	d	Suriname	n
Indonesia	d	Romania	d	Trinidad and Tobago	e
Kiribati	n	Russia	d	Turks and Caicos Islands	n
Korea, Rep.	d	Serbia	d	Uruguay	d
Laos	e	Slovakia	d	Venezuela	d

Key: n no legislation; e enabling legislation only; d detailed provisions.
Sources: EIA Centre (1998); Donnelly *et al.* (1998).

Table 3.2   Features of EA Systems of Selected Countries in Sub-Saharan Africa

Country	Legal requirement for EA	Local and sectoral provisions for EA	Implement- ation	Application to policies and plans	Screening of projects	Scoping	Competent authority for environmental acceptability	Requirements for public participation	Requirements for monitoring	Expertise for conducting EA
Ghana	Environmental Protection Agency Act 490/1994, EIA Procedures 1995	Guidelines issued by EPA for mining sector 1995	Selected projects, sectoral emphasis	No	List with discretionary powers, Preliminary Environmental Review	General checklist, approved Terms of Reference, sectoral guidelines	Environmental Protection Agency	Public notices, optional public hearing	EMP required, EPA monitors	Consultants
Malawi	National Environmental Policy 1996 and draft Environmental Management Act 1996	None	Discretionary Funding agencies	No	Discretionary	No provisions	Director of Environmental Affairs	Policy to make ES public for comment	EMP required	Developer responsible
Mauritius	Environment Protection Act 1991 as amended on 6.4.93	Local and sectoral provisions	Discretionary	No	List, no thresholds	No provisions	Department of Environment	EIS available for public inspection	No formal requirement	Developer responsible
Nigeria	Environmental Impact Assessment Decree 86/1992, Urban and Regional Planning Decree 1992, EIA Procedure 1994	Provisions of Planning Decree	Funding agencies and multinational companies. Approx 55 EIAs submitted and 25 approved 1992–1997	No	List with thresholds, discretionary	Terms of Reference submitted to FEPA for approval, sectoral guidelines for key sectors	Federal Environmental Protection Agency, Federal Environmental Protection Council	FEPA decision published, documents made available, provisions for participation in scoping	Monitoring by FEPA	Consultants
South Africa	Environment Conservation Act 1989 EIA Regulations 1997	Provincial government procedures	Wide voluntary use from 1989, compulsory from 1997	Applied to landuse plans (18% of studies up to 1986)	List of activities	Agreed with competent authority – public participation required	Provincial environmental departments, national for major projects	EIS made public	No formal requirement	Independent consultants

Continued over

Table 3.2 (Continued)

Country	Legal requirement for EA	Local and sectoral provisions for EA	Implement-ation	Application to policies and plans	Screening of projects	Scoping	Competent authority for environmental acceptability	Requirements for public participation	Requirements for monitoring	Expertise for conducting EA
Swaziland	Enabled by Swaziland Environment Authority Act 1992, Environmental Audit, Assessment and Review Regulations 1996	None	Mainly funding agencies. First approval Sept. 1996	No	Illustrative lists	ToR reviewed by SEA	Swaziland Environment Authority	EIA report is a public document	Mitigation plan	Consultants
Tanzania	No general national requirements. Tanzania National Parks Authority guidelines 1993, procedures 1995	Tanzania National Parks, Tanzania Electric Supply Company	24 EIAs 1992–1996	Applies to certain programmes	All developments in National Parks	Checklist	TANAPA (also TANESCO)	Minimal	Monitoring by TANAPA	Consultants, arranged by TANAPA, at fixed cost to developer
Zambia	Environmental Protection and Pollution Control Act 1990, Regulations 1997	None	Little experience	No	Discretionary	Approved Terms of Reference	Environmental Council of Zambia	EIS made public for comment, optional public hearing	Monitoring by developer required	Consultants

Principal sources: Ghana (Allotey and Amoyaw-Osei 1996), Malawi (Luhanga 1996), Mauritius (Barannik and Okaru 1995), Nigeria (Dung-Gwom 1996), South Africa (Hill 1998), Swaziland (Ypma 1996), Tanzania (Mwakilema 1996), Zambia (Government of Zambia 1997).

Table 3.3 *Features of EA Systems of Selected Countries in East Asia*

Country	Legal requirement for EA	Local and sectoral provisions for EA	Implement-ation	Application to policies and plans	Screening of projects	Scoping	Competent authority for environmental acceptability	Requirements for public participation	Requirements for monitoring	Expertise for conducting EA
China	Environmental Protection Law 1979, 1989	Provincial, County and City Environmental Protection Boards, with responsibilities to National Environmental Protection Agency and local government	Tens of '000s per annum (mostly simple). System sometimes by-passed	EIA for regional landuse plans	Simplified and full assessments	No formal requirement	National Environmental Protection Agency, Provincial, County and City EP Boards	No formal provision	Monitoring by local EPA	Licensed organizations
Indonesia	Government regulation No. 29, 1986 Regulation No. 51 of October 1993, Regulations 1996	Sectoral agencies and provincial government procedures	Well established	EIA for regional plans	Lists and screening review	Approved Terms of Reference	Environmental Impact Management Agency (BAPEDAL) and sectoral agencies	Discretionary	Environmental management and monitoring plans	Consultants
Malaysia	Environmental Quality (Amendment) Act 1985 Environmental Quality (Prescribed Activities) Order 1987	EIA process decentralized to DOE State (regional) offices	106 EIAs in 1989, 331 received between April 1988 and December 1991	No	List with thresholds	Preliminary assessment Sixteen sets of specific EIA sector guidelines	Department of Environment and State offices	Limited	Discretionary follow-up by DoE	Registered consultants
Philippines	Presidential Decree No. 1586 of 1978, Order DAO21 1992, Proclamations 1981, 1996, Procedures 1997	Regional offices of Environmental Management Board	In operation for many years	No	Two-stage process, with lists of projects and critical areas	Scoping guidelines	Department of Environment and Natural Resources, Environmental Management Board	Discretionary, optional public hearing	Monitoring provisions	Accredited experts

Continued over

Table 3.3 (Continued)

Country	Legal requirement for EA	Local and sectoral provisions for EA	Implementation	Application to policies and plans	Screening of projects	Scoping	Competent authority for environmental acceptability	Requirements for public participation	Requirements for monitoring	Expertise for conducting EA
South Korea	Environment Preservation Act 1977 EIA Act 1993 (amended 1997), EA Regulation 1993	Local government regulations	1260 EIAs 1981–1994	EA of regional plans	Public and major private projects	General checklist and preliminary EIS	Environmental Assessment Division of Ministry of Environment	Discretionary public hearings	Requirements for post-management	Consultants
Taiwan	EIA Law 1994 Implementation Rules 1995	Provincial Environmental Protection Departments	Between 1985 and 1990 more than 50 EISs were assessed	Regional Development Environmental Assessment, SEA for energy policies	Preliminary study plus general guidance, mandatory list	General list	Local Environmental Protection Bureaux, National Environmental Protection Agency	Public meetings	Yes	State approved experts
Vietnam	Environmental Protection Law 1994, Government Decree 175/CP 1994, guidelines 1994–1997	Delegation to provincial government	300 EIAs 1994 to 1997	No	List, initial environmental examination report, detailed EIA	No formal provision	National Environment Agency and provincial governments	No formal requirement	No formal requirement	Consultant institutes

Principal sources: China (Ortolano 1996), Indonesia (Wen-Shyan Leu et al. 1997), Malaysia (Manp 1998), Philippines (Sajul 1998), South Korea (Han et al. 1996), Taiwan (Wen-Shyan Leu et al. 1997), Vietnam (Friederich 1996).

Table 3.4 Features of EA Systems of Selected Countries in South Asia

Country	Legal requirement for EA	Local and sectoral provisions for EA	Implementation	Application to policies and plans	Screening of projects	Scoping	Competent authority for environmental acceptability	Requirements for public participation	Requirements for monitoring	Expertise for conducting EA
Bangladesh	Enabled by Environmental Protection Act 1995. Draft procedures prepared	EIA guidelines for flood action plan	Primarily funding agencies	No	Green list of exempted projects	No formal provisions	Ministry of Environment and Forestry	No	No	Relevant Ministry
Bhutan	Enabled by National Environmental Protection Act EIA guidelines 1993	None	Primarily funding agencies	No	White, grey and black lists (exempt, partial, full EIA)	Approved ToR	National Environmental Planning Agency	Draft EIA report circulated to concerned parties and NGOs	Provisions for implementation monitoring	Proponent in consultation with NEC
India	Enabled by Environment (Protection) Act 1986 Mandatory under EIA Notification 1994	Limited delegation to State Pollution Control Boards	Implemented for several years	No	List	Sectoral guidelines	Ministry of Environment and Forests	Public hearings made mandatory by 1997 Notification	EMP required. Follow-up by State PCBs	Consultants, mainly national
Nepal	Environmental Protection Law 1996, EP Rules 1997	Sectoral guidelines	Limited experience since rules issued	Ad hoc assessment of Bara Forest management plan	Lists for initial examination and full EIA	Sectoral guidelines, scoping report, approved ToR	Ministry of Population and Environment	EIA report accessible for public review	Audit by MoPE	Consultants
Pakistan	1983 Ordinance No. 37, Environmental Protection Ordinance No. 27 1997 – enabling legislation	Implemented through provincial government	Draft provisions for environmental examination or EIA of PPPs	Limited	Screening guidelines	No provision	Federal and provincial Environmental Protection Agencies	Provision for public review but confidential information withheld	Provisions in draft provincial legislation	Private consultants
Sri Lanka	National Environmental Act No. 47 of 1980 amended 1988, Regulation No. 772/22 1993, Ministerial order 1995	North Western Province statute, powers delegated to sectoral ministries	70 projects submitted in 1996	No	Lists of projects and env. sensitive areas	Guidelines on scoping, scoping meetings	Central Environmental Authority, Project Approving Agencies (sectoral ministries)	EIA reports are open for public review		Private consultants

Principal sources: Bangladesh (Alam 1996), Bhutan (National Environment Commission 1998), India (Modak 1998), Nepal (Uprety 1998), Pakistan (Tunio 1998), Sri Lanka (Yasarante 1998)

Table 3.5 *Features of EA Systems of Selected Countries in Central and Eastern Europe and Central Asia*

Country	Legal requirement for EA	Local and sectoral provisions for EA	Implementation	Application to policies and plans	Screening of projects	Scoping	Competent authority for environmental acceptability	Requirements for public participation	Requirements for monitoring	Expertise for conducting EA
Bulgaria	Environmental Protection Act 1991, Regulation 1992, 1995	Some delegation to regional and municipal authorities	60 national and 1000 local EIAs per year	Mandatory for development plans and landuse plans	List with thresholds	Checklist in Regulations	Ministry of Environment and Waters, Regional Inspectorates, municipal government	Mandatory public hearing. EIS available to public	Yes	Licensed experts
Croatia	Law on Physical Planning 1980, EIS Regulations 1984, Law on Environmental Protection 1994, Decree on EIA 1997	Delegation to local government enabled but not implemented	60 EIAs per year	Enabled by Law, but rarely done in practice	List with thresholds	General checklist, preliminary EIA	State Directorate for the Environment, EIA Commission	Public hearing, EIS available to public	No	Consultants
Estonia	Government Regulation No. 314 1992 Ministry of Environment Regulation No. 8 1994	Delegation to local government	170 EIAs carried out between 1992 and 1995	Plans	Lists and general criteria	Specified by competent authority	Ministry of Environment or the environmental authorities of county/city governments	EIS available to public for comment		Licensed experts selected by competent authority at developer's expense
Latvia	Law on State Ecological Expertise 1990. Law on EIA 1998	None	Large numbers of projects	Physical planning and development programs	Obligatory for all projects		Board of State Environmental Impact Assessment, Regional Environmental Boards	Limited	No	State experts

Country	Legislation	Decentralization	Number	Landuse plans	Screening	Scoping	Central authority	Public participation		Experts
Poland	Environmental Protection Act 1990, Landuse Planning Act 1994, MoE Orders 1995	Delegation to local government if not of national importance	20 per year nationally, many locally	Landuse plans	Lists	Agreed with competent authority	Central Environmental Protection Authority	SEA reports readily available, EIA reports not readily available		Authorized experts
Russia	Environmental Protection Act 1991, Ecological Expertise Act 1995, OVOS Regulations Order No. 222 1994	Delegation to local authorities	Many thousand SERs per year	Plans	No for SER, list for OVOS		State Environmental Review Department, national or provincial	Optional public review, public meetings		Independent experts
Slovakia	Federal Act 17 1992, Act No. 127/1994 on EIA	Participation of municipal government	180 EIAs 1994–1997	Plans	Lists, mandatory, discretionary, thresholds	Defined by Ministry of Environment – general checklist, sectoral guidelines	Sectoral authorities	Public EIS, public hearing	Yes	Authorized experts
Ukraine	Law on Environmental Protection 1991, Law on Environmental Expertise 1995	Sectoral OVOS instructions, local government committees	Many SER	Plans	List for OVOS, no thresholds	Defined by competent authority	Ministry for Environmental Protection and Nuclear Safety	Optional public review, EIS not public		SER state experts

Principal sources: Bulgaria (Veleva and Anachkova 1999), Croatia (Starc, Markovac, and George 1999), Estonia (Peterson 1999), Latvia (Sekakis and Rusza 1999), Poland (Woloszyn 1999), Russia (Cherp 1999), Slovakia (Kozova and George 1999), Ukraine (Patoka 1999)

Table 3.6   *Features of EA Systems of Selected Countries in the Middle East and North Africa*

Country	Legal requirement for EA	Local and sectoral provisions for EA	Implement-ation	Application to policies and plans	Screening of projects	Scoping	Competent authority for environmental acceptability	Requirements for public participation	Requirements for monitoring	Expertise for conducting EA
Egypt	Law 4, 1994, on Protection of the Environment	Coordination with sectoral ministries and local governorates	Initially funding agencies, but expanding	No	Lists for EA required, not required, and preliminary EA	Sectoral guidelines	Egyptian Environmental Affairs Agency, with links to sectoral ministries	EIS not made public	Follow-up by competent authority, linked to pollution control	Local consultants
Jordan	Enabled by Environmental Protection Act 1995	Aqaba Region Directive 1995 (draft)	Funding agencies	No	Lists for EA required, not required, and preliminary EA (draft)	ToR approved by GCEP (draft)	General Corporation for Environmental Protection	EIS not made public	Linked to pollution control	Consultants
Morocco	Enabled by Decree 2-93-809 1994. Environmental Protection Act 1996	None	5 EIAs submitted 1995	No	Lists for mandatory EIA and exempt from EIA	ToR prepared by Ministry	Sectoral ministries	EIS not made public		Consultants
Oman	Environment Protection and Pollution Control Act 1982, amended 1985, 1993	None		No	Discretionary		Administration of Environmental Planning and Permissions, Ministry of Regional Municipalities and Environment	No provision	Temporary licence pending monitoring by officials	Government officials

			Funding agency projects			EIA unit	Sectoral licensing authority		Licensing authority responsible	Central EIA Unit
Syria	None Decree due to come into force 1999	None		No	Lists of mandatory and discretionary projects, optional initial EIA					
Tunisia	Law 88–91, 1988. EIA Decree 91–362, 1991	None	1028 EIAs submitted in 1994	No	Two lists, mandatory and requiring initial evaluation	ToR defined by NEPA	National Environmental Protection Agency	No provision, no public access to EIA report	No	Consultants
Turkey	Environmental Law 1983, EIA Regulations 1993, 1997	None	Well established	EIA of integrated activities enabled	Lists for full EIA and initial EIA	Sector-specific formats defined by MoE	Ministry of Environment, Evaluation and Assessment Commission	Public meeting	Ministry of Environment responsible	Local consultancies

Principal sources: Egypt (Scholl 1996), Jordan (Al-Khoshman 1995), Morocco (Baouendi 1995), Oman (Hewehi 1995), Syria (Ahmad 1996), Tunisia (Baouendi 1995), Turkey (Turkman 1994)

Table 3.7   Features of EA Systems of Selected Countries in Latin America and the Caribbean

Country	Legal requirement for EA	Local and sectoral provisions for EA	Implementation	Application to policies and plans	Screening of projects	Scoping	Competent authority for environmental acceptability	Requirements for public participation	Requirements for monitoring	Expertise for conducting EA
Belize	Environmental Protection Act 1992 Regulations 1995	No		No	List	Guidance in procedures	National Environmental Appraisal Committee	Public review	Procedural requirement	Government officials
Bolivia	Law on Environment 1992, Regulations 1995, 1996	Coordination with local and sectoral government			Screening for comprehensive or specific EIA	Sectoral guidelines	Ministry of Sustainable Development and the Environment		Monitoring plans required	
Brazil	Rio de Janeiro permit system, 1977 National Environmental Law 1981 Executive Decree 88,351 1983 (issued 1986), EIA Regulations 1986	Yes, in many states	225 EIAs carried out up to 1988, of which 91 were in Sao Paulo state	Required in Sao Paolo legislation	List	Guidance	National Council for Environment (CONAMA)	Public hearings. Mandatory publication of EIA	Monitoring required, but limited follow-up	Independent consultants
Chile	Framework Environmental Law 1994 Regulations 1997	Regional Environmental Commissions	186 voluntary EIAs prior to regulations	Local and regional urban zoning plans	List	General checklist	National and Regional Commissions for the Environment	EIS available for comment	Yes	Responsibility of developer

Mexico	Law of Public Works 1980 Federal Law on Environmental Protection 1982 Regulation on EIA 1988 and amendments	State laws and municipal regulations required to be established to support the Federal law	Extensive		Lists for federal and state level		State offices of Social Development Secretariat and state and municipal government	Limited access to documentation	Little follow-up	Certified consultants
Peru	Environment and Natural Resources Code 1990	Sectoral Ministries and Municipality of Lima have issued own regulations	Primarily funding agencies		Lists	General list in code	Individual Ministries, e.g. Energy and Mines, Fisheries Coordination by National Environmental Council	Public access to EIS, public meeting required by Ministry of Energy and Mines	Required by sectoral regulations	Registered bodies
Uruguay	Law of Environmental Impact Assessment 1994	Sectoral ministries responsible for evaluating projects					Ministry of the Environment, sectoral ministries	Discretionary		Proponent responsible
Venezuela	Organic Law of Environment 1976 Regulation on EIS, Decree 2213, 23/4/92	Detailed requirements set by sectoral ministries	About 50 EIA 1970–1992, increasing numbers since 1992	No	Lists of projects and sensitive areas		Ministry of Environment and Renewable Natural Resources. Linked to sectoral ministries and landuse planning	Limited	Limited	Developer responsible

Principal sources: Belize (Government of Belize 1995), Bolivia (Donnelly et al. 1998), Brazil (Brito and Moreira 1995), Chile (Contreras 1996), Mexico (Lomelí and Galaviz 1996), Peru (Aldana 1996), Uruguay (de Mello 1995), Venezuela (Chico 1995)

Table 3.8 *Features of EA Systems in Selected Development Banks and Aid Agencies*

Agency	Procedural requirement	Application to policies and plans	Screening of projects	Scoping	Requirements for public participation	Requirements for monitoring
World Bank	Operational Policy, Procedure and Practice Guide 1999	Guidance for sectoral and regional loans	Guidance lists for full EIA, Environmental Appraisal, none (categories A, B and C)	EA sourcebook and Updates	EIS publicly available (categories A and B)	Monitoring plan required. Provisions for supervision and post-auditing visits
African Development Bank	EA guidelines 1992		Categories I, II and III	Sectoral guidelines	No provisions	Monitoring, supervision and post-evaluation by country and Bank staff
Asian Development Bank	EA procedures 1993		Guidance lists for categories A, B and C (full EIA, initial environmental examination, none)	Sectoral guidance	EIS publicly available if not confidential	Monitoring plan required
European Bank for Reconstruction and Development	Environmental Procedures 1996		List for full EIA, broad guidance for Environmental Analysis	Not defined	Guidance. EIS publicly available if not confidential	Monitoring plan required
Inter-American Development Bank	Procedures 1990		Categories I, II, III and IV			
Australia AUSAID	Guidelines 1996		Categories 1 to 5	Joint screening and scoping exercise	Reports available to the public	Provisions for monitoring and ex-post evaluation
Canada CIDA	Procedural guide 1995		List for comprehensive study			Public access to EIA report
Denmark DANIDA	Procedures 1994	Procedures for sector programme support	Categories A and B for full and partial EIA	Sector checklists	No provisions	Broad monitoring by Bank staff

Agency	Principal documents		Screening/categories	Scoping	Consultation	Monitoring
European Commission DG1A/1B/8	DG1B 1997 DG8 1993		Categories A, B and C, environmental analysis or full EIA	Sectoral checklists and guidelines	Broad guidance	Broad monitoring and evaluation by Commission staff
Finland FINNIDA	Guidelines 1989		Preliminary and detailed EA	Conducted throughout EIA, sectoral guidelines		Follow-up shared with host country
Germany GTZ and KfW	Guidelines 1995		General guidance	Sectoral sourcebook		Monitoring by recipient, verified by agency, post-evaluation
Japan JICA	Environmental Guidelines 1990–1994		Initial environmental examination, pre-EIA, full EIA	Joint screening and scoping exercise, sectoral guidelines		Host country responsible for monitoring
Netherlands DGIDC	Procedures 1993	Includes policies and programmes	Based on EC Directives	Formal scoping process	Formal requirements for consultation and public meetings	Monitoring plan and overall monitoring and evaluation by agency staff
Norway NORAD	Guidelines 1990–96		Initial assessment, full EIA	Sectoral checklists and guidelines		
United Kingdom DFID	Procedures 1996		Environmental Appraisal, full EIA	Sectoral and environmental checklists	Detailed guidance	Broad guidance
USA USAID	Procedures 1980	Applies to plans and programmes	Guidance list and exemption list. Initial environmental examination, full EIA	Scoping procedure	Provisions for public consultation and hearings	Cooperation with host country on monitoring programme

Principal documents: African Development Bank (1992), Asian Development Bank (1993), Australian Agency for International Development (1996), Canadian International Development Agency (1995), Danish International Development Agency (1994), European Bank For Reconstruction and Development (1996), CEc (1993), European Commission (1997), Finnish International Development Agency (1989), Germany: BMZ (1995), Inter-American Development Bank (1990), Japanese International Cooperation Agency (1990–94), Netherlands Directorate-General for International Co-operation (1993), Norwegian Agency for Development Cooperation (1990–96), Overseas Development Agency (1996), United States Agency for International Development (1980), World Bank (1999).

References

African Development Bank (1992) *Environmental Assessment Guidelines*, African Development Bank, Abidjan

Ahmad, B (1996) EIA in Syria and other Arab States; a Comparative Review, unpublished MSc dissertation, University of Manchester

Alam, M (1996) Recent EIA developments in Bangladesh, *EIA Newsletter* **13**: 9

Aldana, MI (1996) Recent EIA developments in Peru, *EIA Newsletter* **13**: 12

Al-Khoshman, M (1995) *The State of Environmental Impact Assessment (EIA) Implementation in the Hashemite Kingdom of Jordan and the Perspectives of Economic Valuation of the EIA Results*, Centre for Environment and Development for the Arab Region and Europe (CEDARE), Cairo, Egypt

Allotey, JA and Amoyaw-Osei, Y (1996) Developing environmental impact assessment procedures for a developing country – the experience of Ghana, *IAIA '96*, Centro Escolar Turistico e Hoteleiro Estoril, Portugal vol. II, pp. 911–914

Asian Development Bank (1993) *Environmental Assessment Requirements and Environmental Review Procedures of the Asian Development Bank*, Asian Development Bank, Manilla

Australian Agency for International Development (1996) *Environmental Assessment Guidelines for Australia's Aid Program*, Australian Agency for International Development, Canberra

Baouendi, A (1995) *EIA and the Role of Economic Valuation in the Arab Countries: Morocco and Tunisia*, Centre for Environment and Development for the Arab Region and Europe (CEDARE), Cairo, Egypt

Barannik, A and Okaru, V (1995) Harmonisation of EA procedures and requirements between the World Bank and borrowing countries, in *Environmental Assessment (EA) in Africa: a World Bank Commitment*, Goodland R *et al.* (eds), World Bank, Washington DC, pp. 35–64

Bellinger, E, Lee, N, George, C, Paduret, A (eds) (1999) *Environmental Assessment in Countries in Transition* Central European University Press, Budapest, Hungary

BMZ (1995) *Environmental Handbook: Documentation on Monitoring and Evaluating Environmental Impacts* Bundesministerium für Wirtschaftliche Zusammenarbeit und Entwicklung, Braunschweig, Germany

Brito, EN and Moreira, IVD (1995) EIA in Brazil, *EIA Newsletter* **11**: 11–12

Canadian International Development Agency (1995) *Procedural Guide to the Canadian Environmental Assessment Act* Canadian International Development Agency, Hull, Quebec, Canada

Cherp, O (1999) Environmental impact assessment in the Russian Federation, in *Environmental Assessment in Countries in Transition*, Bellinger, E, Lee, N, George, C and Paduret, A (eds), Central European University Press, Budapest, Hungary

Chico, I (1995) Environmental management in Venezuela *Environmental Assessment* **3**: 105

Commission of the European Communities (CEC) (1993) *Environmental Manual – Environmental Procedures and Methodology Governing Lomé IV Development Co-operation Projects*, Directorate General for Development, CEC, Brussels

Contreras, L (1996) Recent EIA developments in Chile *EIA Newsletter* **13**: 10

Danish International Development Agency (1994) *Environmental Assesssment for Sustainable Development*, Danish International Development Agency, Copenhagen

de Mello, M (1995) The law of environmental impact assessment, *Environmental Policy and Law* **25**: 73–75

Donnelly, A, Dalal-Clayton, B and Hughes, R (1998) *A Directory of Impact Assessment Guidelines 2nd edn*, International Institute for Environment and Development, London, UK

Dung-Gwom, JY (1996) Environmental Assessments in Nigeria, *EIA Newsletter* **12**: 15–16

EIA Centre (1999) Environmental assessment reference library (http://www.art.man.ac.uk/eia/eiac.htm) EIA Centre, University of Manchester, UK

European Bank For Reconstruction and Development (1996) *Environmental Procedures*, European Bank For Reconstruction and Development, London, UK

European Commission (1997) *Environmental Impact Assessment DG1B Guidance Note*, European Commission, Brussels

Finnish International Development Agency (1989) *Guidelines for Environmental Impact Assessment in Development Assistance*, Finnish International Development Agency, Helsinki

Friederich, H (1996) Recent EIA developments in Vietnam, *EIA Newsletter* **13**: 15

Goodland, R and Edmundson, V (eds) (1994) *Environmental Assessment and Development*, World Bank, Washington DC, USA

Goodland, R, Mercier, J-R and Muntemba, S (eds) (1995) *Environmental Assessment (EA) in Africa: a World Bank Commitment*, World Bank, Washington DC, USA

Government of Belize (1995) *Environmental Impact Assessment Regulations, SI107, 1995*, Department of the Environment, Ministry of Tourism and the Environment, Belize

Government of Zambia (1997) *The Environmental Protection and Pollution Control Regulations 1997*, Government of Zambia, Lusaka, Zambia

Han, E, Kim, M, Lee, J and Kim, S (1996) Environmental impact assessment and environmental assessment *IAIA '96*, Centro Escolar Turistico e Hoteleiro Estoril, Portugal

Hewehi, M (1995) *Environmental Impact Assessment and Economic Valuation in Oman*, Centre for Environment and Development for the Arab Region and Europe (CEDARE) Cairo, Egypt

Hill, R (1998) Environmental Assessment in South Africa, *EIA Newsletter* **16**: 18–19

Inter-American Development Bank (1990) *Procedures for Classifying and Evaluating Environmental Impacts of Bank Operations*, Inter-American Development Bank, Washington DC, USA

Japanese International Cooperation Agency (1990–1994) *Environmental Guidelines* (series), Japanese International Cooperation Agency, Tokyo

Kozova, M and George, C (1999) Environmental impact assessment in the Slovak Republic, in *Environmental Assessment in Countries in Transition*, Bellinger, E, Lee, N, George, C and Paduret, A (eds), Central European University Press, Budapest, Hungary

Lohani, B, Evans, JW, Ludwig, H, Everitt, RR, Carpenter RA and Tu, SL (1997) *Environmental Impact Assessment for Developing Countries in Asia*, Asian Development Bank, Manila, the Philippines

Lomelf, D and Galaviz, JLR (1996) EIA in Mexico: problems and needs, *IAIA '96* vol. II, Centro Escolar Turistico e Hoteleiro Estoril, Portugal

Luhanga, J (1996) Integrating EIA and socio-economic appraisal in the project cycle: the dilemma faced by Malawi, *Integrating Environmental Assessment and Socio-economic Appraisal, 24–25 May 1996*, Development and Project Planning Centre, University of Bradford, UK

Manp, ARA (1998) Malaysia's three step EIA process, *AREAP News Bulletin* 1/1, IUCN Nepal, p. 18

Modak, P (1998) EIA process in tiers *EIA Quarterly* 1, Environmental Management Centre, British Council, Mumbai, India, pp. 4–6

Mwakilema, W (1996) Recent EIA developments in Tanzania – the role of environmental impact assessment (EIA) in protected area, *EIA Newsletter* **13**: 14

National Environment Commission (1998) Bhutan: a brief review of EIA guidelines, *AREAP News Bulletin* 1/1, IUCN Nepal, p. 20

Netherlands Directorate-General for International Co-operation (1993) *Environmental Impact Assessment in Development Co-operation*, Directorate-General for International Co-operation, The Hague, The Netherlands

Norwegian Agency for Development Cooperation (1990–1996) *Environmental Impact Assessment of Development Aid Projects: Initial Environmental Assessments* (series), Norwegian Agency for Development Cooperation, Oslo, Norway

OECD (1996) *Coherence in Environmental Assessment: Practical Guidance on Development Co-operation Projects*, Organisation for Economic Co-operation and Development, Paris, France

Ortolano, L (1996) Influence of institutional arrangements on EIA effectiveness in China, *IAIA '96* Vol. II, Centro Escolar Turistico e Hoteleiro Estoril, Portugal, 901–905

Overseas Development Agency (1996) *Manual of Environmental Appraisal*, Overseas Development Agency, London

Patoka, I (1999) EIA in Ukraine: history and recent developments, in *Environmental Assessment in countries in Transition*, Bellinger, E, Lee, N, George, C and Paduret, A (eds), Central European University Press, Budapest, Hungary

Peterson, K (1999) Upgrading EIA procedures in Estonia, in *Environmental Assessment in countries in Transition*, Bellinger, E, Lee, N, George, C and Paduret, A (eds), Central European University Press, Budapest, Hungary

Sajul, E (1998) Philippines: revising laws for effective enforcement, *AREAP News Bulletin* 1/1, IUCN Nepal, p. 20

Schroll, H (1996) EIA and development in Egypt, *Integrating Environmental Assessment and Socio-economic Appraisal, 24–25 May 1996*, Development and Project Planning Centre, University of Bradford, UK

Sekakis, I and Rusza, S (1999) Environmental impact assessment in Latvia, in *Environmental Assessment in Countries in Transition*, Bellinger, E, Lee, N, George, C and Paduret, A (eds), Central European University Press, Budapest, Hungary

Starc, N, Markovac, Z and George, C (1999) Environmental Assessment in Croatia, in *Environmental Assessment in Countries in Transition*, Bellinger, E, Lee, N, George, C and Paduret, A (eds), Central European University Press, Budapest, Hungary

Tunio, I (1998) Environmental protection in Pakistan, *AREAP News Bulletin* 1/1, IUCN Nepal, p. 19

Turkman, A (1994) Efficiency of environmental impact assessment in the decision-making process in Turkey, in *NATO/CCMS Pilot Study*, report of the Seventh Workshop, University of Antwerp, Belgium, pp. 59–67

United Nations Environment Programme (1995) *UNEP's Way Forward: Environmental Law and Sustainable Development*, United Nations Environment Programme, Nairobi, Kenya

United Nations Environment Programme/United Nations Development Programme (1996–1998) *Compendium of Environmental Laws in African Countries*, UNDP/UNEP, New York, USA

United Nations Environment Programme (1996) *EIA: Issues, Trends and Practice* (prepared by Bisset, R), UNEP, Nairobi, Kenya

United States Agency for International Development (1980) *Environmental Procedures* 22CFR 216 United States Agency for International Development, Washington DC, USA

Uprety, B (1998) Nepal's legal system for EIA *AREAP News Bulletin* 1/1, IUCN, Nepal, p. 18

Veleva, V and Anachkova, S (1999) Environmental Impact Assessment in Bulgaria, in *Environmental Assessment in Countries in Transition*, Bellinger, E, Lee, N, George, C and Paduret, A (eds), Central European University Press, Budapest, Hungary

Wen-Shyan Leu, Williams, WP and Bark, AW (1997) Evaluation of environmental impact assessment in three Southeast Asian nations, *Project Appraisal* **12**: 89–100

Woloszyn, W (1999) Environmental Impact Assessment in Poland, in *Environmental Assessment in Countries in Transition*, Bellinger, E, Lee, N, George, C and Paduret, A (eds), Central European University Press, Budapest, Hungary

World Bank (1999) *Operational Policy, Bank Procedure and Good Practice 4.01: Environmental Assesssment* World Bank, Washington DC, USA

Yasarante, SE (1998) Sri Lanka's legal procedures for EIA *AREAP News Bulletin* 1/1, IUCN Nepal, pp. 21–22

Ypma, P (1996) Recent EIA developments in Swaziland, *EIA Newsletter* **13**: 13–14

4

Screening and Scoping

Christopher Wood

4.1 Introduction

This chapter discusses the principles and practice of screening and scoping in environmental assessment (EA) in developing countries (LDCs) and countries in transition (CITs). The activities involved in these two early stages of the EA process may be defined as follows:

- *Screening*: deciding whether the nature of the action and its likely impacts are such that it should be submitted to environmental assessment
- *Scoping*: determining which environmental impacts should be included in the assessment

Screening involves the assessment of the likely overall significance of the *combined* environmental impacts of the action. Scoping, on the other hand, involves the assessment of the likely significance of *individual* environmental impacts of the action. Both activities require that a working definition of significance be developed.

The United States Council on Environmental Quality (CEQ) Regulations contain ten general criteria for the determination of significance on a case-by-case basis (Box 4.1). These criteria are intended to be applied within the social and environmental context in which the action would occur. Many of the criteria have been adopted or modified for use by other jurisdictions. Decisions about the screening and scoping of projects and other actions in developing countries are based, consciously or unconsciously, on criteria very similar to those in Box 4.1.

Environmental Assessment in Developing and Transitional Countries. Edited by N. Lee and C. George.
© 2000 John Wiley & Sons, Ltd.

Box 4.1   US Council on Environmental Quality Significance Criteria

1. Is the impact adverse or beneficial?

2. Does the action affect public health or safety?

3. Is the action located in a unique geographic area?

4. Are the effects likely to be highly controversial?

5. Does the proposed action pose highly uncertain or unique or unknown risks?

6. Does the action establish a precedent for future actions with significant effects, or represent a decision in principle about future considerations?

7. Is the action related to other activities with individually insignificant but cumulatively significant impacts?

8. To what degree may the action affect designated or listed and protected sites?

9. To what degree may the action adversely affect endangered or threatened species and habitats?

10. Could the action contravene other environmental legislation?

Source: amended from CEQ Regulations 40 CFR 1508.27(b).

This chapter consists of two main parts: Section 4.2 deals with screening and Section 4.3 with scoping. In each case, the main methods in use are reviewed, illustrations are provided of their application in an LDC or CIT context, an evaluation of their relative strengths and weaknesses is made and suggestions for improving their application are advanced. The chapter ends, in Section 4.4, with a number of conclusions.

4.2   Screening

It is important that the effective screening of actions takes place in all EA systems. Without it, unnecessarily large numbers of actions would be assessed and some actions with significant adverse impacts may be overlooked. The determination of whether or not an environmental assessment is to be prepared for a particular action should hinge upon the likely significance of its environmental impacts. Two broad approaches to the identification of such actions may be distinguished:

- The compilation of lists of actions and of accompanying thresholds and criteria (which may include locational characteristics) to help in determining which actions should be assessed
- The establishment of a procedure (which may include the preparation of a preliminary or intermediate environmental assessment report) for the case by case (discretionary) determination of which actions should be assessed

In practice, most EA systems adopt a hybrid approach to screening which involves the use of lists, thresholds and the use of some discretion in the selection of projects.

In some systems, screening also determines which type of EA (with different documentary and participation requirements) is to be undertaken for actions of different size, type or potential environmental significance.

In the case of policies, plans and programmes formal screening for EA purposes is normally undertaken on the basis of lists (e.g. all new land use plans, or major revisions of policies). However, where procedures are less formalized there may be considerable discretion in how screening decisions are made.

To undertake screening a very preliminary assessment of the environmental impacts of the action must be made. For projects, this involves having some basic knowledge about the nature of the proposal, such as its scale, the quantities of wastes to be generated, its appearance and some information about the characteristics of the proposed location. For policies, plans and programmes, similar, but less specific, knowledge about the projects which will implement the higher level decisions is needed.

Where screening is carried out utilising *lists of actions*, these are usually of either a positive or inclusive nature or of a negative or exclusive nature. Thus the World Bank (1999) guidelines include illustrative lists of 'Category A' actions (which may require a full environmental assessment [Box 4.2]) and 'Category C' actions (which are unlikely to require any environmental assessment). Those actions included in Category C cover: education; family planning; health; nutrition; institutional development; and human resources projects. A weakness of lists, which are only based on types of actions, is that individual actions within the same category can vary greatly in size, technology, resource requirements, waste characteristics, physical layout and location and can, therefore, have environmental impacts of widely different significance.

Box 4.2   World Bank 'Category A' Projects/Components

The projects or components included in this list are likely to have adverse impacts that normally warrant classification in Category A.

- Dams and reservoirs
- Forestry production projects
- Industrial plants (large-scale) and industrial estates, including major expansion, rehabilitation, or modification
- Irrigation, drainage, and flood control (large-scale)
- Aquaculture and mariculture (large scale)
- Land clearance and levelling
- Mineral development (including oil and gas)
- Port and harbour development
- Reclamation and new land development
- Resettlement
- River basin development
- Thermal power and hydropower development or expansion
- Manufacture, transportation, and use of pesticides and other hazardous and/or toxic materials
- New construction or major upgrading of highways or rural roads
- Hazardous waste management and disposal

Source: World Bank 1999.

Lists are, therefore, often employed discretionarily (as in the World Bank procedures) or in association with thresholds and/or criteria. Thresholds may relate to such features as the action's size, demand for raw materials, emissions or outputs or the amount of land it requires. However, there is often debate about the appropriateness of particular thresholds. There may, for example, be little difference between the significance of the impacts of a new 10 km road which is subject to EA and a 9.5 km road which falls below the EA threshold. There is also a risk that setting EA thresholds encourages the sub-division of actions so that each component falls below the threshold even though, in combination, they exceed it.

Screening criteria may also include the type of area in which the action will be located. This allows the incorporation of sensitive locations into the screening procedure. Thus, all actions of a certain type (e.g. roads) within certain areas designated as sensitive (e.g. national parks or Ramsar sites) may require an EA irrespective of their size or other characteristics. Similarly, consideration of the carrying capacity of the receiving environment may be appropriate when determining whether particular programmes, plans or policies are to be subject to EA. In South Africa, all projects of certain types and all specified changes of use must be subject to EA, though not necessarily to a full EA (Republic of South Africa 1997). The same applies in the case of 29 categories of project in India (Banham and Brew 1996).

Discretionary screening procedures may contain provisions for a preliminary environmental assessment – essentially a simplified EA report. Once the preliminary EA report has been prepared, it may be accepted as providing sufficient information for decision making purposes. On the other hand, if potentially significant environmental impacts are revealed, a full EA report may then be required. Thus an action may fall into one of four categories: requires no EA; requires a preliminary EA report but not a full EA report; requires a preliminary EA report and then a full EA report; and requires a full EA report.

A number of development funding bodies also make use of preliminary EA reports. These include the European Commission's aid programme for Lomé countries (Commission of the European Communities 1993), which has an 'initial screening' stage, the African Development Bank (1992), the Norwegian aid agency (NORAD 1989) and the UK Department for International Development (Overseas Development Administration 1996). South Africa uses a 'scoping report' for a similar purpose (Republic of South Africa 1997). This two stage approach leads to the preparation of at least two types of EA report in a number of developing countries. This is true in Egypt (Egyptian Environmental Affairs Agency – EEAA 1996), in India (Banham and Brew 1996) and in South Africa (Republic of South Africa 1997).

In practice, the procedures for screening actions often mix different approaches and this permits different levels of scrutiny and of consultation and participation. Of the 25 EA systems examined by Sadler and Verheem (1997), a combination of lists (with or without thresholds) and case-by-case assessment was used in 60% of the cases to identify which actions should be subject to EA.

This *mixed approach* is used in a number of developing countries (Chapter 3). For example, in Egypt three lists of projects and criteria are employed, the 'white list', the 'grey list' and the 'black list'. For white list projects, no environmental impact assessment (EIA) is necessary. For grey list projects a 'scoped EIA study' may be required and black list projects require a 'full EIA study'. Two environmental screening forms are used to report the expected environmental impacts of white list (Form A) and grey list (Form B) projects. The screening procedure in Egypt is shown in Figure 4.1.

In India, the Ministry of Environment and Forests determines whether a 'rapid' (one month) or 'full' (one year) assessment is to be undertaken for the 29 categories of project requiring mandatory assessment. There is also provision for the Ministry to grant a project a specific exemption from any assessment (Banham and Brew 1996). In South Africa, the relevant authority determines whether the scoping report, which must be prepared for all the specified projects, is sufficient for

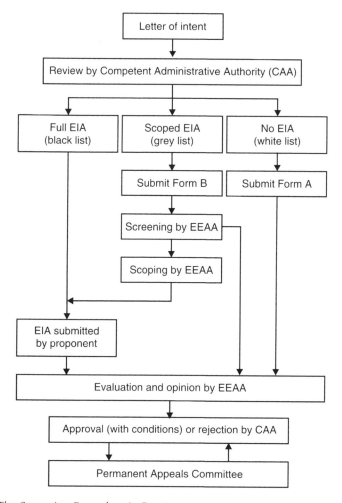

Figure 4.1   *The Screening Procedure in Egypt*

decision-making purposes or whether a full environmental impact assessment must be undertaken (Government of South Africa 1997).

In practice, the screening of actions for the applicability of environmental assessment is often not undertaken satisfactorily in a number of developing countries. Environmental agencies tend to be weak and are not always able to ensure adequate implementation of their own screening procedures. Where overseas development agencies are involved, their own screening procedures (World Bank 1991; Commission of the European Communities 1993) may be stricter and more influential.

In a study of 26 Tanzanian EIA reports, prepared prior to the introduction of a mandatory EA system, Mwalyosi and Hughes (1998, p. 35) reported that:

> In the absence of rigorous screening processes, development proposals in Tanzania are generally subjected to much lower levels of environmental assessment than would be the case if, say, World Bank screening criteria had been used. . . . Of the 26 project EIAs reviewed, only 7 (27%) were actually 'full' and in-depth EIA processes. Had the World Bank's screening criteria been used, 17 (65%) would need to have been subjected to a 'full' EIA process.

Kennedy (1988) has highlighted the need to have a sufficiently simple and effective screening system in place in developing countries to improve their *assessment practice*. This should comprise a list of projects and other actions, together with accompanying criteria and thresholds to determine whether or not an action should be subject to environmental assessment. Some form of discretionary procedure should also be put in place to resolve occasional difficulties of assignment and differences of view. In many cases, this simple system should extend to the use of simplified environmental assessment for appropriate projects. If screening is to be operated effectively, the proponent should be required to submit appropriate information to assist the decision-maker and/or the relevant environmental authorities in determining whether an environmental assessment is necessary in any particular case.

Since proponents of actions require as much certainty as possible in determining whether assessment is likely to be required, sufficiently clear and precise information about actions, criteria, thresholds and screening procedures generally should be available. Such guidance is also helpful to the other participants in the assessment process (see, for example, European Commission 1995). To instill confidence in the process, the screening decision for each action should be made by a publicly accountable body, and the reasons for that decision should be publicly recorded (Wood 1995; 1997).

In countries in transition, screening is often not formally undertaken to restrict the numbers of state ecological reviews to manageable numbers (Bellinger *et al.* 1999). More specific project lists for screening purposes and the specification of SEA screening arrangements have both been recommended, together with the making public of screening decisions (Bellinger *et al.* 1999).

Where a number of overseas funding agencies are engaged in the development process, problems can arise due to inconsistencies between their individual screening requirements and between these and any pre-existing in-country screening procedures which are in force. In this situation, there is a strong case for greater coherence and co-ordination between their screening requirements, preferably based upon commonly accepted criteria of impact significance. Indeed, there is a more general case for greater coherence in procedures at all stages of the EA process between different aid agencies, development banks and in-country developer/environmental organizations (see Chapters 12–14 for further details).

4.3 Scoping

Scoping is the process of determining the range of issues to be analysed in the environmental assessment and then recorded in the assessment report (e.g. the EIS). In other words it is concerned with determining the terms of reference (TOR) of the assessment. Its purpose is to ensure that more focused assessments are carried out and that more relevant and useful EISs are prepared. Scoping has proved to be a valuable innovation and many jurisdictions have now incorporated scoping procedures into their EA systems. Of the 25 EA systems examined by Sadler and Verheem (1997) half had specific scoping requirements and all but two utilized some form of scoping. The majority of these provided for some form of consultation during scoping, involving the environmental authorities and, in some instances, the public. The scoping findings took a number of forms – guidelines for the future assessment were published in less than half of the jurisdictions surveyed and in only 50% of these were the guidelines binding on proponents. Scoping may also take the form of a preliminary EA report referring to feasible alternatives, likely key impacts, gaps in information, the further investigations to be undertaken and the types of assessment methods to be used.

Scoping for the environmental assessment of policies, plans and programmes is just as essential as for the EIA of projects. The scoping process may, however, be more complex. As in project EIA, it is essential to consider the geographical area affected and to refine the list of impacts to those which are likely to be significant for decision-making purposes. As in EIA scoping (discussed below), checklists, matrices and networks, scenario analyses and comparison with similar studies, expert judgement and consultation are frequently used in SEA scoping (Therivel and Partidario 1996).

Scoping commences with the preliminary identification of impacts. There exist numerous *identification methods* that may be used for this purpose (Bisset 1988; Glasson *et al.* 1999; Morris and Therivel 1995; Canter 1996; Ortolano 1997). Those most frequently used are checklists, matrices, consultation and expert opinion; often more than one method is used in combination. Use may also be made of published guidelines, whether of a general or generic nature, or previously pre-pared action-specific guidelines or EA reports (Donnelly *et al.* 1998). Such guidelines may be helpful either because they relate to actions similar to that which is proposed or to actions in similar locations.

Checklists comprise lists of biophysical and other features of the environment and lists of the impact generating characteristics of actions which can be used in the systematic identification and preliminary assessment of the potential impacts of an action. These checklists may be general or more generic in nature (e.g. relating to particular types of environment or types of action). Though useful, checklists are of limited value in identifying impact interactions or secondary and indirect effects. Like all aids to assessment, they need to be employed as a guide, not as a prescriptive 'cook-book' (European Commission 1995b; Ministry for the Environment 1992). Box 4.3 presents an extract from some scoping guidelines for Indonesian wetlands which identify processes associated with the harvesting of plant products which may be unsustainable, the benefits that may be lost or reduced as a consequence, and the types of mitigation measures that may be investigated (Asian Wetland Bureau 1991).

Matrices are essentially two-dimensional check lists. They list the characteristics of the proposed action against the characteristics of the environment and the constituent cells are used to record whether the different elements of the proposal will have impacts on the different characteristics of the environment. The simple matrix is easily understood and can be applied to a wide range of types of action to establish their direct impacts. It is also useful for communicating likely impacts to readers of scoping reports. However, matrices may be large and unwieldy and are not very satisfactory for identifying significant indirect, secondary and cumulative impacts.

An example of a matrix, used in scoping and reporting the environmental impacts of a waste-water sewer project in Egypt, is illustrated in Figure 4.2. It indicates both the wide range of mainly negative impacts which may be associated with the construction phase of the project as well as the mainly positive environmental impacts that may occur during its operational phase (Taylor, Binnie and Partners 1997).

Networks, and simpler cause-effect flow diagrams, are used to help in tracing the web of relationships which exists between the different activities associated with an action and the environmental system with which they interact (Sorensen 1971 (cited in Canter 1996)). These diagrams can be particularly helpful in identifying indirect and cumulative impacts. However, because of their greater complexity they demand greater knowledge and expertise for their effective use.

The impacts identified when assessors use these approaches may still not include some which are potentially important. *Consultations* with decision-making and environmental authorities, with environmental interest groups and with the local community and its representatives, should assist in ensuring that all potentially significant impacts are detected. However, there is also the opposite danger, particularly where extensive consultations takes place, that some non-significant impacts will be unjustifiably included in the TOR for the assessment. The use of expert opinion is possibly the best check on this (Wood 1995).

In *practice*, scoping in many developing countries tends to be more fully developed in those situations where this is a requirement of the aid agency or development bank funding the project. There is a limited form of scoping in Indian EA systems. The proponent of a project usually discusses the content of the EA report with the Ministry of Environment and Forests, but there is normally no public involvement, although general provisions for public participation have recently been strengthened (Banham and Brew 1996). Deficiencies in scoping practice are a contributory factor in the production of EA reports which provide a great deal of data 'but still miss or underplay discussion of the critical issue' (Modak 1998, p. 5).

Box 4.3   Indonesian Wetlands Scoping Guidelines for Plant Product Harvesting

Activities associated with harvesting of plant products include:

- Selective logging
- Harvesting of plant products such as jelutung, fruit, rotan, nipah and nibung

These activities can occur either sustainably, such as where regeneration replaces extracted timber (over long periods), or unsustainably, such as where a combination of logging intensity, soils, slopes and climate reduces the ability of the vegetation to regenerate.

Impacting processes

- Harvesting of products at a rate greater than that at which they naturally grow and regenerate is unsustainable, and will result in the eventual reduction of that species to levels at which harvesting is not worthwhile

In this way many or all of the benefits of that species will be lost. Other species which depend on the harvested species may also be affected. In some cases the harvested species and/or the dependent species may become locally extinct

- Harvesting of products which have a nature conservation value. Any loss of these species is undesirable

Benefits reduced or removed

- Protection from natural forces
- Nutrient removal and/or retention
- Toxicant removal and/or retention
- Source of natural products (on-site)
- Gene bank
- Significance for nature conservation
- Socio-cultural significance
- Research and education site
- Maintenance of existing processes or natural systems
- Good representative of this class of wetland

Mitigating activities

Regulation through quotas for amounts of products taken.

Minimize hydrological disruption when extracting timber, particularly avoiding digging timber extraction canals.

Harvest products sustainably and selectively.

Leave core areas unharvested and undisturbed to act as sources of seed for regeneration.

Leave seed trees at appropriate densities.

Examine alternative sites which may not possess important wetland benefits.

Notes

In many cases, detailed studies are required in order to determine whether or not activities are sustainable.

Source: Asian Wetland Bureau 1991, pp. 273, 274.

	Land use	Topography	Landscape	Aesthetics	Macro-climate	Micro-climate	Air quality	Odours	Ambient dust	Noise	Vibration	Geology	Soil resources	Soil quality and fertility	Groundwater levels	Groundwater quality	Surface water	Flooding	Vegetation cover	Flora and fauna	Land tenure	Future planned development	Cultural and historical sites	Archaeology	Local communities	Indigenous peoples	Population	Employment	Public health	Safety	Traffic flows	Industry	Land value
CONSTRUCTION																																	
Site clearance	-			-					-	-			-	-			-		-		-										-		
Access roads	-	-	-	-			-		-	-	-		-	-			-		-				-							-	-		
Building demolition	-			-					-	-	-		-	-																-	-		-
Earthworks	-			-			-		-	-	-		-	-			-						-	-							-		
Importation of equipment							-			-													-										
Construction materials							-			-	-					-								-				*					
Shaft excavation			-	-					-	-	-	-		-			-							-	-					-			
Grouting																																	
Dewatering																				-													
Tunnelling											-	-																					
Spoil removal and storage	-		-	-				-	-	-										-			-	-	-					-	-		-
Spoil haulage and disposal	-		-	-				-	-	-																					-		
Air compression	-								-	-																							
Site compounds	-						-		-																								
Labour force																						*			*			***	***				
Connections to pumping stations	-																								-			-		-	-		-
Reinstatement	-		*	***										*					*	*					-					-			
OPERATION																																	
Operation of tunnel			***	***				***						*		*	*	***				***	*	*	***			-	***	*			*
Ventilation of tunnel			-	-			-	-						*		*	*	***					*	*	***				*	*			
Drainage alteration				*										*		*	*																
Maintenance requirements																													*	***			
Power consumption						*	*																										
Reconnection to sewer network	***		***											*		*	*	***				***	*	***	***				***				*

Key: ***   Potential major positive effect   ---   Potential major negative effect
           *   Potential minor positive effect   -   Potential minor negative effect

Source: Taylor, Binnie and Partners 1997

Figure 4.2   Maadi (Cairo) Rock Tunnel Scoping Matrix

In Egypt, scoping as such is not a formal requirement, although guidelines exist for the content of EA studies for different types of project (EEAA 1996). Despite this, some Egyptian EA reports involve scoping (see Figure 4.2). In South Africa a scoping report must be produced for all projects specified as requiring environmental assessment. In addition, the applicant may be required to prepare a 'plan of study for scoping' which is reviewed by the relevant authority prior to preparation of the scoping report. These documents are placed in the public domain once a decision on the scoping of the project has been reached and there are provisions for the applicant to ensure that opportunity is given to the public to participate in the scoping process (Republic of South Africa 1997).

It is generally accepted that strengthening scoping is an important means of *improving EA practice* in developing countries (Ahmad and Sammy 1985; Kennedy 1988; Bisset 1992; Organisation for Economic Co-operation and Development 1996). The UK Overseas Development Administration's (1996) *Manual of Environmental Appraisal* stresses the proactive nature of the scoping process and the role of public participation within it. The African Development Bank (1992) recommends the combined use of checklists and supporting questions to assist with scoping. The preparation of scoping guidelines and guidance, and the use of consultation and public participation, in undertaking scoping studies in countries in transition has also been recommended (Bellinger *et al.* 1999).

Scoping has tended to be given greater emphasis in EA systems as these have matured within developed economies. The same is likely to happen in developing countries, and countries in transition, as their EA experience grows.

4.4   Conclusions

Amongst others, the World Bank (1991), the Organisation for Economic Co-operation and Development (1996), the Overseas Development Administration (1996), the Commission of the European Communities (1993) and the United Nations Environment Programme (1996) have all produced valuable guides to EA which contain useful advice relating to screening and scoping. In many ways, however, UNEP (1988) enunciates the simplest and most important general principles for good EA practice, which also have specific relevance to screening and scoping:

1. Focus on the main issues.
2. Involve the appropriate persons and groups.
3. Link information to decisions about the project.
4. Present clear options for the mitigation of impacts and for sound environmental management.
5. Provide information in a form useful to the decision-makers.

Methodologies, techniques and standards should be selected with the observance of these general principles in mind. At the same time, it is important that the methods used should be appropriate to local circumstances. Biswas (1992, p. 240) has blamed the use of inappropriate, imported methodologies for EA reports which are 'too academic, bureaucratic, mechanistic and voluminous'.

Three general features of the LDC and CIT situation need to be borne in mind when deciding on appropriate screening and, particularly, scoping methods:

1. Environmental conditions in tropical or near-tropical areas render many environmental assumptions, models and standards, derived in temperate zones, inappropriate.
2. Baseline socio-economic and environmental data may be inaccurate, difficult to obtain or seriously incomplete in many situations.
3. The significance attached to particular environmental impacts may be quite different (much less or much greater) between many developing countries and developed countries, due to major differences in cultural, social and economic conditions.

Central to addressing all of these differences and complexities is the assembling of an interdisciplinary assessment team, which includes sufficient indigenous experts augmented, where necessary, with some external support. Wherever possible local people and interest groups should be involved in screening and scoping where they can help not only in determining which actions should be assessed, and in identifying potentially significant impacts, but also in providing relevant baseline environmental data. External experts may be better able to access data held by an over-protective agency than internal ones, but local experts will usually be better judges of the appropriateness of the available data, scoping methods and the significance of impacts.

As with other aspects of EA, there is a need to focus on capacity building and training requirements to improve screening and scoping. The development assistance agencies are turning their attention to the need to help in strengthening the institutional framework for administering EA and local centres of EA expertise to improve actual practice. Many commentators have stressed the importance of training to increase the human resource capacity to undertake and review EAs (Ahmad and Sammy 1985; Biswas 1992; Wilbanks *et al.* 1993). The United Nations Environment Programme (1996) has produced a training manual which provides an excellent resource for such training initiatives. There is considerable agreement that training should be provided primarily within the LDC or CIT concerned and that it should relate not just to central government officials but also to regional administrators and to personnel in environmental consultancies and research institutes (Wood 1995; 1997).

Discussion Questions

1. What are the main weaknesses in screening procedures in developing countries and countries in transition and how could they be strengthened?
2. What are the principal impediments to involving the public in screening and scoping decisions in developing countries and countries in transition?

3. How could scoping procedures in developing countries and countries in transition be improved?
4. How do screening and scoping procedures differ between the EIA of projects and the SEA of policies, plans and programmes in developing countries and countries in transition?

Further Reading

Ahmad and Sammy (1995) and Biswas and Agarwala (1992) deal with screening and scoping procedures in EA in developing countries. Bellinger *et al.* (1999) describe screening and scoping procedures in countries in transition. African Development Bank (1992), Commission of the European Communities (1993), Norwegian Agency for Development Co-operation (1989), Organisation for Economic Co-operation and Development (1996), Overseas Development Administration (1996), United Nations Environment Programme (1988) and World Bank (1991) provide guidance on screening and/or scoping in developing countries. Banham and Brew (1996) (India) and Mwalyosi and Hughes (1998) (Tanzania) provide examples of, *inter alia*, the screening and scoping situation in particular developing countries. Bisset (1992), Biswas (1992), Kennedy (1988), Ortolano (1997), Wilbanks *et al.* (1993) and Wood (1995) provide overviews of, *inter alia*, screening and scoping in developing countries generally.

References

African Development Bank (1992) *Environmental Assessment Guidelines*, African Development Bank, Abidjan, Nigeria
Ahmad, YJ and Sammy, GK (1985) *Guidelines to Environmental Impact Assessment in Developing Countries*, Hodder and Stoughton, London
Asian Wetland Bureau (1991) *Manual of Guidelines for Scoping EIA in Indonesian Wetlands*, AWB and Directorate General of Forest Protection and Nature Conservation, Bogor, Indonesia
Banham, W and Brew, D (1996) A review of the development of environmental impact assessment in India, *Project Appraisal* **11**: 195–202
Bellinger, E, Lee, N, George, C and Paduret, A (eds) (1999) *Environmental Assessment in Countries in Transition*, Central European University Press, Budapest, Hungary
Bisset, R (1988) Developments in EIA methods, in *Environmental Impact Assessment: Theory and Practice*, Wathern, P (ed.), Unwin Hyman, London
Bisset, R (1992) Devising an effective environmental assessment system for a developing country: the case of the Turks and Caicos Islands, in *Environmental Impact Assessment for Developing Countries*, Biswas, AK and Agarwala, SBC (eds), Butterworth-Heinemann, Oxford
Biswas, AK (1992) Summary and recommendations, in *Environmental Impact Assessment for Developing Countries*, Biswas, AK and Agarwala, SBC (eds), Butterworth-Heinemann, Oxford
Canter, LW (1996) *Environmental Impact Assessment* 2nd edn, McGraw Hill, New York
Commission of the European Communities (1993) *Environmental Manual – Environmental Procedures and Methodology Governing Lomé IV Development Co-operation Projects*, Directorate General for Development, CEC, Brussels
Donnelly, A, Dalal-Clayton, B and Hughes, R (1998) *A Directory of Impact Assessment Guidelines* 2nd edn, International Institute for Environment and Development, London

Egyptian Environmental Affairs Agency (1996) *Guidelines for Egyptian Environmental Impact Assessment*, EEAA, Cairo

European Commission (1995a) *Environmental Impact Assessment: Guidance on Screening*, Directorate General for Environment, Nuclear Safety and Civil Protection, EC, Brussels

European Commission (1995b) *Scoping in Environmental Impact Assessment – a Practical Guide*, Directorate General for Environment, Nuclear Safety and Civil Protection, EC, Brussels

Glasson, J, Therivel, R and Chadwick, A (1999) *Introduction to Environmental Impact Assessment*, 2nd edn, University College London Press, London

Kennedy, WV (1988) Environmental impact assessment and bilateral development aid: an overview, in *Environmental Impact Assessment: Theory and Practice*, Wathern, P (ed.), Unwin Hyman, London

Ministry for the Environment (1992) *Scoping of Environmental Effects: a Guide to Scoping and Public Review Methods in Environmental Assessment*, MfE, Wellington

Modak, P (1998) EIA process in tiers – part 1, *EIA Quarterly* (British Council, Bombay) **2**: 4–6

Morris, P and Therivel, R (1995) *Methods of Environmental Impact Assessment*, University College London Press, London

Mwalyosi, R and Hughes, R (1998) *The Performance of EIA in Tanzania: an Assessment*, Environmental Planning Issues 14, International Institute for Environment and Development, London

Norwegian Agency for Development Co-operation (1989) *Environmental Impact Assessment (EIA) of Development Aid Projects – Checklists for Initial Screening of Projects*, NORAD, Oslo

Organisation for Economic Co-operation and Development (1996) *Coherence in Environmental Assessment: Practical Guidance on Development Co-operation Projects*, Development Assistance Committee, OECD, Paris

Ortolano, L (1997) *Environmental Regulation and Impact Assessment*, John Wiley, New York

Overseas Development Administration (1996) *Manual of Environmental Appraisal*, ODA, London

Republic of South Africa (1997) *Environmental Impact Assessment Regulations (Environment Conservation Act 1989)*, Nos R1182–1184, Government Gazette, 387, No18621, 5 September 1997

Sadler, B and Verheem, R (1997) *Country Status Reports on Environmental Impact Assessment: Results of an International Survey*, EIA Commission, Utrecht

Taylor, Binnie and Partners (1997) *Greater Cairo Wastewater Project: Maadi Rock Tunnel: Draft Environmental Impact Statement*, Ministry of Housing, Utilities and Urban Communities, Cairo

Therivel, R and Partidario, MR (eds) (1996) *The Practice of Strategic Environmental Assessment*, Earthscan, London

United Nations Environment Programme (1988) *Environmental Impact Assessment: Basic Procedures for Developing Countries*, Regional Office for Asia and the Pacific, UNEP, Bangkok

United Nations Environment Programme (1996) *Environmental Impact Assessment Training Resource Manual*, Environment and Economic Unit, UNEP, Nairobi

Wilbanks, TJ, Hunsaker, DB Jr, Petrich CH and Wright, SB (1993) Potential to transfer the US NEPA experience in developing countries, in *Environmental Analysis: the NEPA Experience* Hildebrand, SG and Cannon, JB (eds), Lewis, Boca Raton, FL

Wood, C (1995) *Environmental Impact Assessment – a Comparative Review*, Longman, Harlow

Wood, CM (1997) What has NEPA wrought abroad? in *Environmental Policy and NEPA: Past, Present and Future*, Clark, ER and Canter, LW (eds), St Lucie Press, Boca Raton, FL

World Bank (1991) *Environmental Assessment Sourcebook*, Environment Department, World Bank, Washington, DC (3 volumes, with subsequent updates)

World Bank (1999) *Good Practices: Environmental Assessment*, Operational Manual, GP 4.01 Environment Department, World Bank, Washington, DC

5

Environmental Impact Prediction and Evaluation

Clive George

5.1 Introduction

Predicting the magnitude of a development's likely impacts, and evaluating their significance, is the core of the environmental assessment process. It requires specialist technical skills and a thorough understanding of the receiving environment. For projects with diverse impacts, a multi-disciplinary team may be needed, covering many different specialisms. The aims of this chapter are not to provide specialist guidance on predicting individual types of impact, but:

- To describe the general principles of impact prediction and evaluation, appropriate to all disciplines
- To give a broad overview of some of the main techniques used

Developing countries and countries in transition may differ significantly from high income countries, and from each other, in characteristics which affect impact prediction and evaluation. Climate, ecology, geology, population density, social structure and many other factors can vary considerably between countries. The chapter gives a broad indication of some of the differences which may need to be taken into account.

Sections 5.2 and 5.3 describe the general principles of impact prediction and evaluation, and the collection of data for prediction purposes. Sections 5.4–5.8 review prediction techniques for particular types of impact. Section 5.9 examines prediction methods used in strategic environmental assessment, and Section 5.10

Environmental Assessment in Developing and Transitional Countries. Edited by N. Lee and C. George.
© 2000 John Wiley & Sons, Ltd.

considers prediction practice and future trends. More detailed information on specialist techniques can be found in the sources referenced.

5.2   Principles of Impact Prediction and Evaluation

Whatever the impact, and whatever specific technique is used to analyse it, prediction and evaluation should be based on a sound methodological framework, which covers:

- The overall prediction and evaluation process (Box 5.1)
- Choice of prediction technique (Box 5.2)
- Criteria for evaluating significance (Box 5.3)
- The design of mitigation measures (Box 5.4)
- Indirect impacts, long range impacts and uncertainty (Box 5.5)

Overall Prediction and Evaluation Process

Box 5.1 summarizes the basic steps which need to be taken in assessing any impact. Baseline data, including anticipated future changes, need to be gathered for impact identification as part of scoping (Chapter 4). More detailed information may need to be gathered subsequently in order to predict impact magnitudes and evaluate significance (Section 5.3). It is important to distinguish between impact magnitude and impact significance. A given level of noise may have very different significance according to whether it is in an industrial area or a residential area. The significance of air emissions will vary according to whether the baseline air quality is well within ambient standards, or approaching them, or already exceeds them.

In the first place, the magnitude and significance of potential impacts need not be evaluated with any greater precision than is needed for the design of mitigation measures. The magnitude and significance of any residual impact may need to be evaluated more thoroughly.

Box 5.1   The Overall Prediction and Evaluation Process

1. Define the baseline environment as it currently is, in sufficient depth to allow impacts to be identified (but no further).
2. Determine future changes to the baseline in the absence of the action.
3. Define the action in sufficient detail to understand its consequences.
4. Identify the likely significant impacts of the action (scoping).
5. Predict the magnitude of each impact, with sufficient precision to evaluate its significance, and to define mitigation measures (gathering more data if needed).
6. Define mitigation measures to reduce significance.
7. Predict the magnitude of residual impacts.
8. Evaluate the significance of residual impacts.

In order to predict impact magnitude, the magnitude of the *primary effect* must first be determined, such as the quantity of a pollutant emitted, the area of land-take, or the size and appearance of a building. The *connections or pathways* between source and receptors (flora, fauna or human beings) must then be defined, and where appropriate modelled. Once these connections are understood, the *impact on receptors* can be predicted.

Choice of Prediction Techniques

For any impact, a number of different types of technique are generally available (Ministry of Housing, Spatial Planning and the Environment 1984). These are summarized in Box 5.2, together with some of the strengths and weaknesses of each, and guidance on their validation. Weaknesses can often be compensated by using a combination of techniques.

The choice of technique should be appropriate to the circumstances. A highly sophisticated mathematical model requiring extensive baseline data may be totally unnecessary if a rough and ready estimate shows the impact to be insignificant with a high degree of certainty, or if it allows effective mitigation measures to be defined. This may often be the case with small, low cost projects. However, major projects can have major impacts, which often need to be predicted with the same degree of sophistication as would be the case in high income countries. Shortages of data and/or expertise can often create difficulties in low income countries, but if simpler prediction techniques and worst case assumptions are not sufficient to show that an impact is insignificant, extra data may have to be gathered, and/or expertise brought in from overseas.

In countries whose EA systems are new, with limited experience of prediction methodologies, strong reliance is often placed on the last of the techniques summarized in Box 5.2; the professional opinion of the developer's chosen experts. Partly for this reason, the EA systems in many low and middle-income countries require some form of official accreditation of assessors. However, a broad-brush quality control of this nature is unlikely to be as effective as the more specific review techniques discussed in Chapter 8. With more rigorous review systems, the quality of each assessment can be judged on its own merits, and the relevance of each assessor's scientific and technical expertise may be evaluated according to the needs of each component of each particular assessment.

For many types of impact, one or more of the other techniques listed in Box 5.2 will be applicable, and professional experts can be expected to make full use of them. Nonetheless, for particularly complex types of impact, including many ecological ones, there may be no effective model, no complete test that is practicable, and no fully representative experience of similar developments in similar environments. Prediction then does have to rely on expert professional judgement. This should be based on a thorough understanding of the scientific principles and processes involved, familiarity with the particular characteristics of the receiving environment, and extensive relevant practical knowledge. The reasoning behind any professional judgement should always be described, and not just stated as an opinion. For further substantiation, each assessor's academic and professional qualifications and experience should be stated in the EIA report.

Box 5.2   Types of Prediction Technique

	Strengths	*Weaknesses*	*Validation*
Past experience	Particularly valuable for complex effects which cannot easily be modelled, and might not otherwise be identified	Can be unrepresentative of the action being assessed	Actual experiences should be quoted, and allowance should be made for the different characteristics of the proposal and its environment
Numerical calculations or models	Can deal with circumstances which are specific to the action being assessed; quantification of primary effects (e.g. area of landtake) is often straightforward	Use of more complex models requires a detailed understanding of the science, and may require considerable data; hidden errors can arise from inappropriate assumptions and approximations in models	Complex models should only be used when simpler ones are inadequate for the purpose; data sources should be identified and shown to be valid; the validity of the model should be demonstrated, e.g. by referring to relevant professional literature
Experiments or tests	Can model complex effects, e.g. by measuring the noise emitted from machinery, or the effect of a pollutant on a particular species	Can be expensive; may not be fully representative of the action being assessed	Experimental arrangements should be shown to be representative of the proposal
Physical or visual simulations and maps	Useful for visual and other spatial impacts; e.g. physical models, photo-montages, computer graphic images, overlay maps	Misleading if not modelled accurately	Written descriptions may be needed to support the simulation, e.g. in relation to different vantage points, or time-dependency of the impact
Professional judgement	Versatile and easy to apply	Misleading if expertise not adequate for the task; difficult to substantiate	Reasoning and supporting data should be described, and qualifications and experience of each professional should be given in the EIA report

Evaluation of Significance

Once the magnitude of an impact has been predicted, its significance must be evaluated. It is through an appropriate choice of significance criteria that sustainable development objectives can be achieved (George 1998, 1999). However, any divergence between the significance criteria used in EA, and those implicit in a country's political culture and socio-economic frameworks, can result in the whole EA process being subverted and becoming ineffective. Three main forms of significance criteria are defined in the United States EA system (Canter 1996), and are summarized in Box 5.3. Although these are relevant generally, their interpretation can differ quite considerably in low and middle income countries, compared with high income ones.

Box 5.3 Criteria for Evaluating Impact Significance		
	Form of criterion	*Nature of impact significance*
Institutional recognition	Legal requirement or other institutional norm, e.g. policy statement, official guidance, environmental standards	Widely understood and agreed
Public recognition	Opposition to the impact, controversy over it, or conflict between different sections of the community	Not widely enough agreed to have resulted in an institutional norm, but of concern to at least some sections of the public
Technical recognition	Concern based on technical understanding of the impact's consequences	Not widely enough understood to cause public concern, but of concern to technical specialists

Institutional recognition is generally the most important form of criterion in all countries. In addition to specific legal requirements (including international agreements and conventions as well as national laws and standards), relevant policy statements or development plans should be consulted. National or international guidelines are also valuable indicators of significance, even though they may not have full legal force.

Environmental standards can take the form of ambient standards (for the quality of the receiving environment), pollution standards (permissible emission and discharge levels) and technical standards (design factors), all of which should be adhered to. Such standards can be particularly important in countries where EIA serves as a pollution control mechanism as well as a development planning instrument. If national standards have not been defined, international recommendations such as those issued by the World Health Organization (e.g. WHO 1993) may provide useful guidance, or standards issued in similar countries (e.g. CEU 1995).

Public recognition often receives less attention in countries whose democratic processes are not well developed. This can sometimes lead to opposition being underestimated, with consequent difficulties and delays in project implementation. Aid agencies and development banks generally encourage the public's views to be obtained and taken into account (Chapters 9 and 14).

Technical recognition may often come into play for complex impacts which threaten critical ecosystems or resources, but it can be important for any type of impact. In countries without strong democratic traditions, it is often given more importance than public recognition, and may be the main mechanism by which social impacts are evaluated.

Under each of these forms of recognition, the context of the impact should be considered, as well as its intensity or magnitude. An impact of a given magnitude is more likely to be significant if it affects a large geographical area, or a large number of people, or a particularly sensitive location, or if it is of long rather than short duration, or if it is irreversible.

Design of Mitigation Measures

Box 5.4 summarizes the main approaches to mitigation. Compensation arrangements are particularly important for projects such as major dams, which can involve relocation of large numbers of people. These arrangements may not always be considered satisfactory by the people affected, particularly in countries whose democratic processes are weak. Social impact assessment (Chapter 7) and stakeholder participation (Chapter 9) can help to address this.

Box 5.4   Alternative Means of Mitigation

Approach	Examples
Avoid	Change of route or site details, to avoid important ecological or archaeological features
Replace	Regenerate similar habitat of equivalent ecological value in a different location
Reduce	Filters, precipitators, waste water treatment, noise barriers, dust enclosures, visual screening, wildlife corridors, changed timing of activities
Restore	Site restoration after mineral extraction
Compensate	Relocation of displaced communities, facilities for affected communities, financial compensation for affected individuals, recreational park or other compensating environmental benefit

---

The developer's commitment to all proposed mitigation measures should be demonstrated in the EA report, and may be reinforced by approval conditions, environmental management plans and monitoring (Chapter 11).

Additional Factors in Impact Prediction

The prediction and evaluation methodology summarized in Boxes 5.1–5.4 should be applied to secondary and long range effects as well as to immediate impacts, and uncertainty in prediction should also be considered. Box 5.5 summarizes these additional factors.

Box 5.5   Additional Aspects of Impact Prediction

1. Impacts under abnormal environmental or operating conditions.

2. Indirect, higher order and synergistic effects.

3. Cumulative and multi-source impacts.

4. Transboundary and global impacts.

5. Uncertainty.

Abnormal conditions include equipment failures, human errors, accidents, and unusual or abnormal operating or environmental conditions (Section 5.8). *Indirect or higher order effects* are particularly common in the case of ecological impacts and impacts on human beings, but can apply to any type of impact. *Synergistic effects* may occur when different pollutants or other impacts interact with each other, chemically or otherwise, to produce additional impacts. Flowcharts and network diagrams may be useful for identifying such effects. Also, communication, questioning and brainstorming, among all members of the assessment or evaluation team, can often help to identify impacts that might otherwise be overlooked.

For *cumulative and multi-source impacts*, the impact of the action has to be added to those from other past, present and future actions, before its significance can be evaluated. In some cases it may be necessary to undertake complex multi-source predictions, particularly in countries where there is no separate pollution control mechanism in existence.

Transboundary impacts are a special case in the evaluation of significance, because normal decision-making processes do not cater for them. Europe's Espoo convention (UNECE 1991), which has been ratified by several transitional countries as well as other European countries, provides a model for dealing with such impacts. It includes provisions for information exchange, public participation in the affected countries, and resolution of disputes. The outcome of these processes may give an indication of the significance of individual impacts. Where relevant bilateral agreements have been negotiated for particular types of impact, or regional agreements such as the Danube Convention or the Mekong Agreement, these may provide a firm basis for evaluating significance.

Global impacts present additional difficulties, because of their cumulative nature. It is not sufficient simply to show that an impact is small compared with the cumulative total, since this ignores the overall impact's accumulating significance (George 1997). In principle, international agreements are the most reliable means of evaluating significance, but few of these contain specific requirements. The detailed requirements of the Rio climate convention and its Kyoto amendments apply primarily to high income countries. For some projects in low and middle income countries, development banks and aid agencies may require mitigation as a condition of funding. Otherwise, more subjective criteria may have to be used. For example, it may be sufficient to show that the cumulative per capita contribution to global biodiversity loss or greenhouse gas concentration, from the country as a whole over its entire development history, including the additional contribution from the proposed action, will remain less than that which has accumulated from activities in high income countries. This does not stop the effect happening, but it may show that its significance is no greater than has been deemed acceptable by high income countries as part of their own development.

Uncertainty in prediction arises from such sources as lack of accurate data, lack of understanding of the behaviour of complex systems, lack of knowledge of organisms' responses, or assumptions and approximations in models. All sources of uncertainty should be acknowledged, and an indication should be given of the consequent degree of uncertainty in the prediction. Various techniques of ecological risk assessment have been devised (Hope 1995; Gabocy and Ross 1998), but are not easy to apply to complex systems. If worst case assumptions overestimate the impact to the extent that it becomes unacceptable, further data gathering or more detailed analysis may be necessary. Alternatively, a monitoring programme with appropriate contingency plans may be undertaken.

5.3   Data Collection

In countries where EA is relatively new, and assessment experience is limited, EIA reports often contain voluminous descriptions of the baseline environment and the project, which may bear little relation to the report's much briefer sections on impact identification and prediction. This is particularly so when prediction is based primarily on expert opinion. By adopting a more rigorous approach to impact prediction, data needs can be defined more clearly, and the whole EA process can be made more efficient and effective.

Beyond the fairly general needs of impact identification or scoping, the collection of data should be focused specifically on the needs of the prediction techniques to be used, as shown in Box 5.6.

Box 5.6   Obtaining Data for Impact Prediction

1. Define the prediction technique.

2. Define the data needed for prediction.

3. Obtain existing data.

4. Review validity of existing data.

5. Identify gaps.

6. Obtain new data.

Future changes to the environment in the absence of the proposed development should always be taken into account. These may arise from, for example, economic or social changes, climatic or geographical effects, and other expected actions. Seasonal variations must also be taken into account, and can be particularly significant in some of the climates common in low and middle income countries. If existing data are scarce, it may sometimes be necessary to gather information over a full year (as required in India – see Chapter 3).

In general, data are required which relate to the primary effect at source, the connection between source and receptors, and the response of receptors. The first will most commonly come from information on the action, the second from data on the receiving environment, and the third from published information on receptor behaviour (or relevant standards).

Data on the Environment

In low and middle income countries, less existing (secondary) environmental data may have been accumulated than in high income countries. Some of it may be hard to access because it is not in the public domain, and sources of available data may be poorly referenced. However, such sources usually exist, and a large part of the assessor's skill lies in knowing or finding out who has what data. Typical sources include national and local government bodies, technical and academic institutions, libraries, NGOs, other interest groups, and individual members of the public. The data obtained should be checked for whether it may be out of date, and for the reliability of its source. Where there are gaps, field research should be restricted to what is actually needed for prediction purposes. If a simple analysis, based on readily obtainable data, is sufficient to show that an impact is insignificant or to define appropriate mitigation, data collection need extend no further.

Aerial photography and satellite imaging are widely used in low and middle income countries, as they can be a highly cost effective means of gathering some types of data. Geographical information systems have been promoted strongly by some donor agencies (World Bank 1993a, 1995b), and can be useful for storing, accessing and analysing data. Despite such techniques however, on-the-ground field research is normally essential, even if this is restricted to a general survey of land usage and habitat.

Data on the Development

Wherever possible, impact prediction and evaluation should be conducted in parallel with project design, so that alternatives can be evaluated, and mitigation measures built into the design. As a result, design details may not be available when the assessment is started. However, predictions can often be made on the basis of worst case assumptions. If these lead to significant impacts, objectives can then be defined for the detailed design, or for the incorporation of mitigation measures into the design.

The confidentiality of project information can often be an important factor in developing countries and countries in transition. Even for private sector projects, government authorities may be sympathetic to developers' claims that all information is confidential. This makes it particularly important to identify information which is actually needed for impact prediction, not only to facilitate the prediction, but to ensure that the full assessment report can be made available to the public. It is often unnecessary for the environmental assessor to know the full technical details of the proposal, or to publish them, so long as its relevant inputs and outputs are defined.

5.4 Impacts on Water

In both the freshwater and marine environment, impacts on water quality and water quantity may both need to be assessed. Impacts on water quality may result from chemical, biological and thermal pollution. Erosion and siltation can affect both quality and quantity, and can have consequential impacts on soils and eco-systems. Quantity impacts include increase/decrease in surface water and ground-water flows and availability, changes in groundwater level, flood risks, and risk of drought. Consequential impacts may include waterlogging and/or salination of agri-cultural land, damage to benthic species or wetland ecosystems, and other effects on ecology and human health, including the creation of breeding sites for mos-quitoes and other disease carriers.

Industrial, agricultural, transport (including ports) and urban developments (in-cluding waste treatment and disposal) are all potential sources of water pollution. Impact prediction generally entails quantifying pollutants at source, modelling their transport, and evaluating the ecological or human health impacts on receptors. Source quantities may be predicted from mass balance calculations or specific tests, and/or from a risk assessment (Section 5.8). Techniques for modelling transport from source to receptor are broadly similar to those for modelling water quantity effects (see below). Additionally, however, the assessment may need to consider mixing, dispersion, solution, adsorption onto silts, and accumulation as sediments, of soluble pollutants, insoluble liquids and particulates, and their interaction with vegetation and soils. All of these effects can be modelled, but reliable data may not be readily available. Professional judgement based on past experience, worst case assumptions and simple calculations may often be sufficient. Pollutant concentra-tions received by receptors should be added to baseline values, for comparison with water quality standards. If no appropriate standards have been established, an assessment of the impact on ecology and/or human health may need to be carried out (Sections 5.7 and 5.8).

Quantity effects may be caused directly by abstraction, storage, channelling, and other direct impacts on flows. They may also occur from changes in landuse such as forest clearance, which can cause major changes to the amount of precipitation infiltrated into the soil or transpired back into the atmosphere, and hence the amount of run-off and erosion. Consequent changes in sedimentation may affect water flows, and must be taken into account in predicting flood risks. Large dams can have far-ranging upstream and downstream impacts on groundwater, so that extensive hydrological data may be needed for prediction purposes. Other data needed for prediction may include land cover and soil properties, precipitation, flows and currents, covering full seasonal variations and extreme climatic conditions.

Some effects such as erosion can be estimated from published data for different types of material and cover, while run-off, surface water and groundwater flows, currents, and sediment transport and deposition can be modelled mathematically. Computerized models are available for this, but complex models may require more data than are readily available. A combination of simple modelling and professional judgement, based on worst case assumptions, past experience and a knowledge of local conditions, may often be the more reliable approach.

Further information on prediction of water impacts, including references on particular modelling techniques, is given in Allison and Durand (1991), Canter (1996), Lumbers (1985), O'Sullivan (1994), Patera and Riha (1996), and World Bank (1995a).

5.5   Impacts on Air and Noise

Air

Prediction of air pollution is similar in principle to prediction of water pollution. Source emissions may be predicted from published process emission factors, a mass balance calculation, or specific tests. Fugitive emissions (including dust) may need to be estimated from past experience or published data.

A rough estimate of pollutant concentration levels can be obtained from a simple box model, which assumes that the pollutant disperses uniformly into an identifiable windstream, bounded on either side according to the source characteristics and terrain, and above by a clearly demarcated stable layer. More accurate approximations model the actual dispersion, most commonly using simple Gaussian equations or more sophisticated fluid dynamics methods (Box 5.7). Various equations have been defined for gases and for particulates, and for different source characteristics, land types, terrain, and climatic conditions. Many of the most widely used of these have been developed or recommended by the US Environmental Protection Agency (USEPA). The simpler versions can be calculated by hand, but the ready availability of low cost computer programs running on personal computers generally makes this unnecessary. Although these programs are easy to use, a good understanding of the science is needed for them to be applied correctly, as the choice of model and input data can have a major influence on the results.

Different calculations may need to be performed in order to assess short-term peaks and long-term cumulative effects. Concentrations are normally predicted at the locations of all critical receptors, both ecological and human. These may include the closest human habitation, nearby large populations, and the closest sites of ecological importance. Concentrations need to be added to the baseline values, including any expected future changes, in order to evaluate their significance in relation to ambient standards or receptor behaviour.

Long range effects, including the development's contribution to general air pollution levels and greenhouse gas concentration, are predicted in much the same way

---

Box 5.7   Thermal Power Stations in India – Air Pollution Impacts

The power stations at Bidadi Hobli (300 MW combined cycle gas turbine) and Talcher (500 MW coal fired extension) are typical medium scale projects in India's programme to satisfy increasing demand for electricity. Both assessments of impact on air quality gathered baseline data on small particulate matter (SPM), SO_2, NO_x, wind speed and wind direction over three months, from three sampling stations at Talcher and 17 at Bidadi. Pollutant quantities emitted were determined from measured fuel compositions and estimated process efficiency factors, and the flue gas and stack data were taken from design documents. The Talcher assessment was undertaken before computers became readily available in India, and the calculations were done by hand using a standard double Gaussian point source equation, with graphical computation of stability values. The Bidadi study used a computer-based fluid dynamics model incorporating ground level turbulence and 3-dimensional modelling of ground terrain. Neither EIA report quotes any validating tests on the models used, nor indicates the degree of uncertainty in the predictions. In both cases calculations were carried out for a variety of wind speeds and directions and meteorological conditions, using the baseline data gathered, and meteorological records. Best case, worst case and average ground level concentrations were calculated, and added to measured baseline data for comparison with ambient air quality standards.

References: Transoft (1997) *Badadi Hobli Rapid Environmental Impact Assessment Report* Transoft, Bangalore, India; Water and Power Consultancy Services (1991) *Rapid Environmental Impact Assessment Study for Talcher Thermal Power Station Extension* Water and Power Consultancy Services, New Delhi, India.

---

as shorter range impacts, but using different dispersion models. In this case the evaluation of significance needs to take particular account of the cumulative nature of the impact.

Further information on predicting air impacts is given in Canter (1996), Cernuschi and Giugliano (1992) and Marriott (1997a), including references to the most commonly used dispersion models.

Noise

Noise can be significant during construction, and also in the operation of many types of development. These may relate to mining, quarrying, roads, railways, and various industrial installations. Since noise is rarely constant, different standards may be specified for fairly steady noise, short sudden noises, or general noise. A widely used standard is for LA_{10}, the A-weighted level in decibels that is exceeded for 10% of the time (A-weighting adjusts for the frequency-sensitivity of the human ear). Standards in residential areas may be lower than in industrial areas, while road traffic noise may have to comply with specific standards. Noise emitted at source is usually determined either from published data for similar types of emitter, manufacturers' specifications, or from specific measurements. Attenuation with distance follows an inverse square law, but is also strongly affected by the direction of radiation from the source, reflection, absorption, wind, temperature gradients, and atmospheric conditions. A number of calculation methods have been defined to take account of these factors, for different types of source, and are incorporated into various computer models. As with all such models, the choice of model and input data can have a significant effect on the results, and in some countries the calculation methods are specified in noise standards.

Once the noise levels at receptors have been calculated, these must be combined with baseline values for comparison with standards, using appropriate formulae to allow for differing noise characteristics and the logarithmic nature of the decibel scale. Noise standards generally relate to the impact on human beings. Effects on fauna, such as disturbance due to sudden noise, may need to be evaluated from experience recorded in the professional literature. Further information and comprehensive reference lists are given in Ortolano (1997), Canter (1996) and Therivel (1995).

5.6   Impacts on Soils, Geology and Natural Resources

Soils

Impacts on soils fall into three main groups; loss of fertility, toxicity, and erosion and its consequences. Any of these may be indirect results of air or water impacts, for example through erosion, pollution, waterlogging, salinization, and changes in land use and run-off. Fertility can also be lost through removal of topsoil, compaction (from vehicles, storage areas, overploughing or overgrazing), and disruption of soil structure or of drainage systems. Toxicity may be generated indirectly via air or water pollution impacts, from industrial and agricultural projects, or by disturbance of buried toxins. Erosion by water or exposure to wind may be caused by any change in land use, and can itself affect fertility. In the semi-arid regions common in many developing countries, loss of fertility can lead to desertification. In tropical climates laterization (the accumulation of metal oxides through tropical weathering) may be another important impact. The consequential impacts of erosion may include damage to wetlands or marine ecosystems such as corals, through the deposition of sediments, and the impacts of dust on human health.

Although many of the effects can in principle be modelled mathematically, some of the effects can be complex, and impact prediction is often largely qualitative. A baseline study of the soil characteristics, and a knowledge of the activities that will take place during construction and operation, allows the likely impacts to be identified, based on experience of similar activities in similar environments. Appropriate mitigation measures can then be designed, based on the effectiveness of previous mitigation experience.

Geology

The principal geological impacts are depletion of mineral resources, loss of or damage to scientifically important geological features or fossils, risk of subsidence, and seismic risk.

The depletion of minerals can be calculated directly, and compared with the cumulative depletion rate of known reserves and estimated total resources. Effects on scientifically important features are of their nature qualitative, and prediction requires a knowledge of the geology and of the activities that will take place. Subsidence can occur as a result of any project involving underground extraction. Medium or large scale extraction projects are likely to require geologists in the design team, and the subsidence risk should then be assessed as part of the design process. The cumulative risk of subsidence from small scale projects may be evaluated by undertaking a geological appraisal as part of the planning process. Seismic activity can be an important factor in the risk assessment of hazardous installations (Section 5.8), and can also be triggered by water resources projects (Box 5.8). Further information on predicting impacts on soils and geology is given in Hodson (1995), Marriott (1997b) and World Bank (1997d).

Box 5.8   Xiaolangdi Multi-purpose Dam – Seismic Impact

The Xiaolangdi dam on China's Yellow River is 154 m high and 1677 m long, and will create a 75 mile long reservoir. Two of its most significant potential impacts are the social effects of resettlement (see Box 5.10) and the potential for triggering seismic activity. The region has a history of serious earthquakes, and several faults were identified in the reservoir area. To establish the baseline, the area was mapped geologically, including the location, length and attitude of faults, and rock samples were tested. Past experience was reviewed by conducting a review of reservoir-induced earthquakes within China and throughout the world. A detailed numerical analysis was carried out to give estimates of earthquake probability and magnitude. However, the amount of quantitative and descriptive information available was limited, and prediction had to rely largely on expert judgement. Expert groups of geologists and seismologists were assembled to discuss the issues, and their conclusions were debated with a panel of international experts. The stability of the dam, embankments and intake towers were then modelled under postulated earthquake conditions of epicentre location and peak acceleration, to assess the risk of failure. Conservative figures were used in the design of structures. In view of the residual uncertainties in prediction, it was recommended that a network of seismicity monitoring stations be established. It was concluded that small earthquakes in some areas were almost certain and detailed predictions were used in planning resettlement and in the preparation of earthquake response plans.

Reference: Yellow River Conservancy Commission (1993) *Xiaolangdi Multipurpose Dam Project Environmental Impact Assessment* World Bank report E0030, CIPM Yellow River Joint Venture, Quebec.

5.7   Impacts on Ecology

Typical causes of ecological impact are loss of habitat, habitat fragmentation, habitat damage, change in drainage conditions, disturbance of species, and the effects of pollutants on species. To predict pollution effects, the pathways between source and receptor must be analysed, which entails close co-ordination with the prediction of impacts on air, water and soils, and a good basic understanding of the relevant ecosystems.

In general, prediction aims to identify any reduction in the abundance or health of any species which may be affected, directly or indirectly. However, because ecosystem behaviour is often very complex, detailed prediction can be extremely difficult. Few models exist, and prediction generally relies on past experience and professional judgement.

For some types of impact (e.g. landtake), the prediction can often be restricted to the first order impact on habitat. If the site has a conservation status, or if rare or endangered species have been identified in the baseline survey, it may be necessary to show that the impact on habitat is insignificant or zero. If the habitat is a common one, and if the area affected is an insignificant proportion of the total area in the region, it may be inferred that the overall effect on species is insignificant. However, cumulative effects may need to be taken into account, particularly in areas rich in biodiversity.

Where the impact on habitat is significant, it may be necessary to consider the impact on individual species that have been identified. Fragmentation effects for example, including loss of connecting corridors, may be particularly significant for particular species. Drainage changes may also have particular impacts on individual species, as will disturbance, and the effects of pollutants. Although it may still not be practicable to quantify the impact on abundance or health of a species, a thorough professional knowledge can provide at least a broad indication of impact magnitude.

Having predicted an impact on one species (positive as well as negative), it may be necessary to predict any consequential impacts on others. Eutrophication is an example of effects in which an impact that is beneficial for certain receptor species (aquatic plants and phytoplankton) can have a catastrophic impact on an entire ecosystem. Flow charts and network diagrams can be a useful aid for predicting consequential impacts, but must be based on a thorough understanding of the affected ecosystem. Local expertise in the particular country, and often in the particular area, is normally essential.

Further information on predicting ecological impacts is given in Hope (1995), Treweek (1995), Morris (1995), World Bank (1995a) and World Bank (1997b).

5.8   Impacts on Human Beings

Socio-economic and cultural impacts are covered in Chapters 6 and 7. Other potential impacts on human beings are health impacts, including those which may occur under accident conditions, impacts on architecture, archaeology and historic sites, and visual impacts.

Health Impact Assessment

A large part of the significance of impacts on water, air, noise and soils arises from their potential impacts on human health. In many cases it is not necessary for EIA to go beyond a comparison with relevant environmental standards, since the consequential health impacts should have been taken into account in defining the standards.

Communicable diseases, non-communicable diseases, malnutrition, injury and mental health (arising from stress conditions) are the principal categories of impact

---

Box 5.9   Oil Exploration off the Namibian Coast – Ecological Impacts

The environmental assessment of Chevron's exploratory oil drilling in Area 2815 off the southern Namibian coast was conducted under the company's petroleum agreement with the Namibian government and the company's own environmental policies. An earlier impact study was carried out prior to seismic surveys, followed by a scoping study for exploration drilling. This led to the preparation of 16 specialist studies covering the baseline environment, trajectory modelling for oil spills, and toxicities. These were reported in a companion document to the EIA report. Impacts on the intertidal zone, wetland habitats and birds, coastal birds, pelagic seabirds, seals, whales and dolphins, and commercially important seaweeds, lobsters, oysters and fisheries were assessed, and also the impact on terrestrial ecology likely to occur from clean-up operations. Predictions were made for normal operations, for the worst case oil spill in the most likely drilling area, and the worst case spill in the worst case drilling area. Detailed data were insufficient for numerical analysis, and so prediction and evaluation were based on professional judgement, using a panel of experts with a wide range of expertise, whose findings were reviewed by independent referees. Impacts were classified as high (highly significant locally, or significant regionally or internationally), moderate (local), low or zero. Each prediction was also given a certainty rating, based on the degree of information and scientific understanding. Mitigation measures were defined for all highly or moderately significant impacts, and for all low significance ones with high uncertainty. All impacts were judged to be of low significance after mitigation, although uncertainty remained high in some cases.

Reference: CSIR-Environmental Services (1994) *Environmental Assessment for Exploration Drilling in Offshore Area 2815, Namibia* CSIR-Environmental Services, Stellenbosch, South Africa.

---

to be considered. All of these may be important in low and middle income countries, where the consequences of poverty can make people particularly susceptible. The incidence of communicable diseases such as malaria may increase as a result of providing new habitats for vectors and carriers, for example due to water resources developments. Other diseases may be introduced by migrant workers and camp followers. Non-communicable diseases which may need to be assessed include those from pollution-related effects for which there is no relevant standard. Injury can be a major hazard for construction workers, operating staff, and in some circumstances, local communities, particularly under accident conditions. Malnutrition and stress problems can result from, for example, poorly planned resettlement programmes. If significant impacts of this nature are likely, an environmental health professional should be involved.

In predicting impacts, the levels of immunity and vulnerability of different sections of the community need to be taken into account, and also the extent of existing health services and infrastructure (e.g. water supply, sanitation, refuse disposal). Where the infrastructure is weak, impacts can be higher than they otherwise would be.

Pollution-related effects can be predicted by calculating pollution levels (Sections 5.4 and 5.5), estimating the doses likely to be received by each section of the affected community, and applying a dose–response relationship, to estimate the likely incidence of disease. However, where dose–response relationships have been documented, they are often derived from data in high income countries, and in many cases there is little reliable local information. Quantification is also difficult with most other types of health impact. Broad estimates of risk may have to be made, based on an understanding of the effects and experience of similar activities in similar situations. The main aim of the assessment is to identify potential impacts, and design mitigation measures through which they can be avoided. Where significant uncertainty remains, it may be necessary to set up a health monitoring programme. Mitigation may often include infrastructural improvements and the provision of clinics, through which impacts can be monitored and remedial action taken if necessary. Health statistics are often scarce in low and middle income countries, and so any monitoring programme may need to be started early enough to define the baseline.

Further information on health impact assessment is given in World Bank (1997a), Birley (1995) and Asian Development Bank (1992).

Risk Assessment

Accidents can have serious health and other impacts on human beings. Many large-scale engineering developments such as chemical works or large dams will require a full risk assessment, separate from the environmental assessment, carried out to standards specified by the appropriate regulatory agencies and/or professional bodies. Since many issues are common to both assessments, environmental assessors should liaise with the risk assessors. For developments not subject to such requirements, the environmental impacts under abnormal conditions need to be assessed within the environmental assessment. Although a full risk assessment may not be necessary for most developments, similar principles apply.

Impacts under abnormal conditions may well be considerably higher than under normal conditions, and their acceptability depends partly on how likely it is that such conditions will actually occur, and with what frequency. A thorough assessment will entail:

- Identifying all potential hazards
- Assessing the environmental consequences should a hazardous situation occur
- Assessing the likelihood, or risk, of the situation occurring

Hazards may typically be caused by mechanical or technical failures, human errors, failure of external supplies such as water or electricity, or external influences such as earthquakes, transport impacts or sabotage. Hazard identification can be carried out by unstructured brainstorming, or by structured techniques which either begin with an analysis of the development and work through to possible hazards, or begin with possible hazards and work back to possible causes. These techniques include hazard and operability studies (HazOps), failure mode effects analysis (FMEA), fault trees, event trees, task analysis, incident analysis, and cause and effect analysis. Some hazards are quite obvious, but others can result from very complex sequences of interlinked events, so that one false assumption can lead to a major hazard being ignored.

Once a hazard and the sequence of events leading to it have been identified, the probability of each event may be estimated, and the overall risk calculated. Quantification of the risk depends on the accuracy of the probability data. When this is poor, worst case assumptions need to be made. In some instances only a very broad estimate of risk may be feasible. Even then, some indication of risk magnitude is necessary, such as 'infrequent', or 'unlikely during the life of the development'.

The likely impact on health in the event of an accident is evaluated in the same way as for operational pollution impacts, by estimating doses received and applying dose–response relationships. This is then multiplied by the risk of the accident, to give the overall health risk. There is no such thing as a 'standard' for what level of risk is acceptable, but for involuntary risks that are life threatening, some indication can be derived from natural risks, such as being struck by lightning. It is generally considered that a lifetime risk of death of less than one in a million is trivial, while a risk of over one in 10 000 is high. Somewhat higher voluntary risks may be tolerated, for example, by people whose jobs depend on the development. It is, however, important to ensure that the people involved understand the risks they may be taking.

In some cases it is instructive to compare the potential cost of taking a risk with the cost of avoiding it, or with the cost of achieving similar health improvements by other means. In low income countries, relatively small per capita expenditures on health care can give much larger benefits than in high income countries. Similarly, improvements in standards of living arising from the development can sometimes have a bigger positive impact on health than the direct negative impacts. Some major infrastructure projects in developing countries (Box 5.10) are designed specifically to reduce risks to health or loss of life. Further information on risk assessment is given in Allison and Durand (1991), Asian Development Bank (1990), Carpenter (1995) and World Bank (1997c).

Architecture, Archaeology and Historic Sites

Potential impacts on sites of special interest can be identified from archaeological and other maps and records. These sites may not be well documented in low and middle income countries, but consultation with local communities may provide additional information. Where the presence of important archaeological sites is suspected, a field survey may be necessary, and monitoring procedures can be introduced during construction, with an archaeologist available on call. Prediction entails estimating the likelihood that unidentified archaeological features may be destroyed during excavation, and predicting possible damage due to construction activities, vibration, drying out, and erosion from weathering or from vehicular and pedestrian traffic. Visual impacts may also be important (see below). Some of these effects can be evaluated numerically, while others rely on past experience and professional judgement. The effects of air pollution can be evaluated from pollutant concentrations and the susceptibility of the materials. The significance of impacts may be determined on the basis of national classifications of important sites and relics, or on professional evaluation.

Box 5.10   Xiaolangdi Multi-purpose Dam – Impacts on Human Beings

The prime purpose of China's Xiaolangdi dam is to control the flooding of the Yellow River, which has claimed millions of lives over thousands of years. The river rises about 10 cm per year as a result of sedimentation, and is already 10 m above the level of the surrounding plain. The dam will trap part of the sediment and facilitate flushing to the sea. It was predicted, principally from past experience, that the alternative of continuing to raise the dykes would result in increasing risks of dyke failure, and the likely loss of many more thousands of lives. A corresponding risk assessment was undertaken for the proposed dam (Box 5.8). Other alternatives considered included planting of trees and crops to reduce erosion, but much of the catchment area is too dry for this to be practicable. One of the aims of the project is to provide irrigation water for this area, and hence achieve a long term solution.

The reservoir will displace some 180 000 people, which is made possible by using irrigation water from the project to bring new land into cultivation. Prediction of social, economic, cultural and health impacts was based largely on past experience of similar resettlement projects, using a wide variety of specialist expertise. Quantified predictions were made of the expected increase in standard of living for the affected communities, based on estimates of increased agricultural productivity. The social assessment included prediction of effects on income from both farming and other sources, housing, amenities, public services and infrastructure, ethnic conflicts, community cohesion, cultural and archaeological heritage and landscape. The prediction of potential health impacts included specialist assessment of the vulnerability of groups such as the very old and the very young, the sick and disabled, women, and minority groups. Evaluation of the significance of potential and expected residual impacts included public consultation on the resettlement plan. This entailed interviews with about 5% of households in both the resettler and host communities, access to resettlement offices, and numerous discussions with local government officials at village and higher levels. Uncertainties in prediction were taken into consideration through proposals for a monitoring programme, including periodic surveys and interviews with the affected population. A detailed set of compensation principles and criteria was drawn up, and a special support fund was set aside to guarantee minimum incomes. A grievances and complaints mechanism was established, to be handled in the first instance by village committees or community associations, and referable through the local government structure to provincial level.

Reference: Yellow River Conservancy Commission (1993) *Xiaolangdi Multipurpose Dam Project Environmental Impact Assessment* World Bank report E0030, CIPM Yellow River Joint Venture, Quebec.

Visual Impacts

Visual impacts can be predicted using physical models, drawings, photo-montages, or computer-graphic simulations. Landscape architects have devised various techniques for classifying and even quantifying visual effects, which go beyond defining the size of an object and the distance and angles from which it can be seen. However, visual representations of one form or another, supported by explanatory descriptions, are the most common form of prediction. So long as they are accurate representations, they enable the significance of an impact to be assessed directly by all those involved in the consultation, participation and decision-making processes. Further information on cultural, archaeological and visual impacts is given in Landscape Institute/Institute of Environmental Assessment (1995), Radmall (1992), Stamps (1997) and World Bank (1994b).

5.9   Prediction and Evaluation in Strategic Level Assessments

The general methodological framework outlined in Boxes 5.1–5.6 can apply to prediction at the strategic level of policies, plans and programmes (PPPs) as well as to project level assessment. The principal differences in prediction at the strategic and project levels are:

- The level of detail required for strategic assessment is normally less
- Greater emphasis is usually placed on predicting the aggregate/cumulative effects of constituent or consequential developments of the PPP as a whole
- Greater emphasis is often placed on the evaluation of alternative PPPs
- The prediction of impacts arising from constituent or consequential developments is likely to be based on their typical rather than their specific features (which are not normally known early in the planning or policy making process)

Box 5.11 summarizes a methodology for prediction in SEA which addresses these differences. For some types of action, such as a structural adjustment programme, the consequential developments which may be expected to occur as a result of implementing a policy can only be defined in very broad terms. For others, such as a coastal development strategy (Box 5.12), they may be defined more precisely, often as a result of past experience. In other cases, such as the development of a water resources plan, it might be practicable to define several alternative plans to the full level of detail of all the individual developments associated with each. However, this may often not be necessary, if it is possible to select the best strategic alternative from a more general analysis, so that fully detailed assessment can be deferred to subsequent project level assessments.

Box 5.11   SEA Prediction Methodology

1. Identify alternative options for the PPP, including the no change or no action option.

2. Identify all forms of constituent or consequential development which may result from each option.

3. Identify typical features of such developments.

4. Identify the typical impacts of such developments, and predict their possible cumulative magnitude and significance.

5. Identify potential mitigation measures.

6. Predict the effectiveness of mitigation measures in reducing impacts, and any other impacts they may have.

7. Evaluate alternatives, taking account of social and economic as well as environmental considerations, and make recommendations.

8. Establish a monitoring programme and contingency plans for possible unmitigated impacts.

Box 5.12   Strategic Impact Management of Egyptian Coastal Development

The development of tourism on Egypt's Red Sea coast includes many small and medium scale activities. These can have considerable cumulative impacts, particularly on the coral reefs, which are ecologically important, and are a major component of the area's touristic attractiveness. In order that the tourist and other coastal industries could be developed sustainably, the Egyptian Environmental Affairs Agency (EEAA) co-ordinated a strategic environmental study of the region's development, with assistance from Denmark and the Netherlands. Prediction was based primarily on past experience of the impact of actual developments. Various forms of damage to corals were identified, from construction activities, increased erosion and consequent sedimentation, discharge of pollution, and mooring of boats on the reefs. Similar impacts were predicted to occur if existing policies remained unchanged. Expert judgement was used to predict the likely impacts of alternative policies, to select the most suitable on environmental and economic grounds, and to identify appropriate mitigation measures. With the support of the Tourist Development Authority, the EEAA issued a set of development guidelines incorporating these policies and measures. The guidelines provide a simple means of checking the environmental acceptability of small and medium scale developments, and can also simplify the assessment of larger projects requiring individual EIAs. Factors covered include the control of erosion and sedimentation, the determination of building set-back lines, the design of marinas, embankments and jetties, the management of water sports, the designation of public beaches, restrictions on hotel ships, measures controlling non-tourist developments, and provision of infrastructure facilities. Developers are required to maintain monitoring records for inspection by the competent authorities, who may themselves gain access for unannounced inspections.

References: National Committee for Integrated Coastal Zone Management (1996) *Integrated Coastal Zone Management in Egypt: towards an Egyptian Framework ICZM programme* Egyptian Environmental Affairs Agency, Cairo; Egyptian Environmental Affairs Agency (1996) *Environmental Guidelines for Development in Coastal Areas* Egyptian Environmental Affairs Agency, Cairo.

As with project level EIA, a distinction needs to be drawn between impact identification (scoping), and prediction of impact magnitude and significance. If potential impacts can be readily mitigated, it may be unnecessary to go much further than impact identification, using methods such as checklists, matrices, cause and effect diagrams, scenario analysis, Delphi techniques (which co-ordinate judgements made by relevant experts), or spatial analysis (e.g. using a Geographical Information System). However, if residual impacts are likely to be significant, estimates of likely magnitude and significance may need to be made.

When an existing policy or plan is being assessed, or one for which there are precedents, the most reliable prediction technique may be inference from past experience. Any or all of the other techniques summarized in Box 5.2 may be used where there is no such experience, or to predict the effectiveness of alternative PPPs or alternative mitigation measures, and any additional impacts they may have. For some types of action, such as industrial zone development plans, water resource plans or transport plans, recourse to sophisticated multi-source dispersion modelling, catchment area modelling, transport sector modelling, or other complex techniques may be beneficial, to reduce reliance on worst case assumptions and restrictive mitigation measures. For other types of action and impact, professional judgement may be adequate, based on sound reasoning.

The degree of uncertainty in predictions depends on the degree of precision with which constituent or consequential developments can be defined. Where the details of these developments are highly uncertain, the assessment becomes one of predicting the cumulative magnitude and significance of impacts which might occur, as opposed to those which can definitely be expected to occur. In some cases, it may be necessary to evaluate the risk of significant impacts occuring. Alternatively, worst case assumptions may be made. If it cannot readily be shown that the risk is acceptably low, appropriate mitigation measures may need to be defined. These will often be policy statements or design principles for consituent or consequential developments, which prevent the effect happening, or which only come into effect if it does.

When uncertainty remains high, which is often the case with strategic level assessments, a monitoring programme may be needed, with appropriate contingency plans. Where it is appropriate to set targets (e.g. for ambient pollution levels), interim targets may be defined within the time period of the PPP, so that corrective action may be taken if monitoring shows that interim targets have not been met.

Strategic environmental assessment is often an integral part of the policy making or planning process itself, within which the significance of environmental impacts needs to be judged alongside social and economic considerations, in order to determine the best overall alternative. Various techniques may be used, such as economic valuation of environmental effects (Chapter 6), social impact assessment (Chapter 7), cost-effectiveness analysis, cost-benefit analysis and multi-criteria analysis (Chapter 10), together with wide consultation and public participation (Chapter 9), to aid the decision-making process.

Further information on prediction in strategic environmental assessment is given in Kozova *et al.* (1996), Canter and Sadler (1997) and World Bank (1993b, 1994a, 1995a and 1996b).

5.10 Prediction Practice and Future Directions

Internationally, the areas of prediction and evaluation practice that are the least well supported by established techniques are in strategic assessment, the assessment of complex ecological effects and risks, the evaluation of transboundary and global impacts, and the integration of environmental, social and economic factors (UNEP 1996; Sadler 1996; Canter and Sadler 1997). Assessments in low and middle income countries can be expected to benefit from improvements in all of these areas. However, as discussed in Chapter 8, prediction practice is one of the weakest areas of environmental assessment, even where established techniques exist. This was the case in the early years of implementing EA in high income countries, and remains so for many assessments in low and middle income countries. There is often excessive reliance on professional opinion, with little use of more objective techniques or substantiating arguments.

Shortage of available data on the environment has been cited as a serious difficulty in making satisfactory predictions in low and middle income countries. This does not correspond with the research findings reported in Chapter 8, which show that description of the environment is undertaken to a satisfactory standard in many assessment reports. Shortage of scientific and technical expertise has also been cited, but assessments carried out for funding agencies often make use of local expertise, to produce assessments that are thorough enough to meet the agencies' quality criteria.

Prediction practice is to a large extent conditioned by the requirements of competent authorities. In countries whose EA systems are fairly new, there is often a lack of consultancy organizations with EA expertise and experience. EIA reports may initially provide little more than baseline data, project data, and the consultants' professional opinions on likely impacts. Competent authorities may compensate for this with a review process which is itself based on professional judgements, made by the reviewers, using the baseline data and project data provided in the report, and any additional data they may call for. In consequence, a review process may become established which accepts inadequate prediction practice, and allows it to continue.

The specialist scientific and technical expertise necessary for impact prediction often exists in low and middle income countries, but primarily in academic and technical institutions rather than in consultancy organizations. In order to improve prediction practice, competent authorities have to establish a framework which encourages this expertise to be tapped, and which encourages developers and consultants to build up their own expertise. Guidance is needed for developers, on how to set about choosing consultants who are capable of managing the assessment process. Guidelines are needed for consultants, on what types of predictions are required, and on where the necessary specialist expertise might be obtained. Universities, technical institutions and sectoral authorities can all make major contributions to developing such guidelines. With these in place, objective review processes can be established, which do not rely solely on the reviewers' professional judgement, but which evaluate the prediction methods used in the EA study itself. Opportunities may then be created, through which consultancy organizations can gain the necessary experience in conducting adequate studies. Such opportunities may for example arise from working with experienced consultants on overseas funded projects, or from networking with professionals in other countries in the same region.

Education and training also have important roles to play. However, EA training itself can do little to teach impact prediction skills, which cannot be learned in a few weeks. The main role of EA training is to provide expertise in identifying what specialist skills are needed, and in how to make use of them within the EA process.

It should also be recognized that detailed science-based impact prediction can be time-consuming and expensive. Indirectly, prediction practice depends on screening and scoping systems which focus the available expertise on those actions and those impacts which most need it. More fundamentally, however, good prediction practice depends on having a review system which accepts nothing less.

Discussion Questions

1. In what ways may impact prediction in low and middle income countries differ from prediction in high income countries?

2. What factors influence the level of detail necessary for data gathering and impact prediction? What factors influence the extent to which impacts can and should be quantified?
3. What information is needed in an environmental assessment report to give confidence that predictions have been carried out thoroughly and professionally?
4. What information is needed in an environmental assessment report to give confidence that impact significance has been evaluated objectively?

Further Reading

General methodologies of impact prediction, together with particular methods for a wide range of specific impacts, are described and discussed in Canter (1996), Morris and Therivel (1995), Marriott (1997), Canter and Sadler (1997) and Ministry of Housing, Spatial Planning and the Environment (1984). A worldwide review of prediction practice is given in Sadler (1996), and recent trends are discussed in UNEP (1996). The World Bank's series of Environmental Assessment Sourcebook Updates (World Bank 1993a,b; 1994a,b; 1995a,b; 1996a,b; 1997a–d) summarize prediction methods for several types of impact, particularly in low and middle income countries, and also for agency funded strategic assessments. Examples of strategic assessment methodologies are described in IUCN Nepal (1995), Kozova *et al.* (1996), National Committee for Integrated Coastal Zone Management (1996) and Tanzania National Parks (1993, 1994).

References

Allison, RC and Durand, R (1991) Problems of risk assessment in water resources management, *The Environmental Professional* **13**: 326–330

Asian Development Bank (1992) *Guidelines for the Health Impact Assessment of Development Projects,* Environment Paper No. 11, Asian Development Bank, Manila

Asian Development Bank (1990) *Environmental Risk Assessment: Dealing with Uncertainty in Environmental Impact Assessment,* Environment Unit, Asian Development Bank, Manila

Birley, MH (1995) *The Health Impact Assessment of Development Projects,* Overseas Development Administration, HMSO, London

Canter, LW (1996) *Environmental Impact Assessment,* 2nd edn, McGraw Hill, New York

Canter, L and Sadler, B (1997) *A Tool Kit for Effective EIA – Review of Methods and Perspectives on their Application,* Environmental and Groundwater Institute, University of Oklahoma, Norman

Carpenter, RA (1995) Risk assessment, *Impact Assessment* **13**: 153–187

Central European University (CEU) (1995) *EBRD/EC Phare/CEU Environmental Standards Database,* Environmental Sciences and Policy Department, Central European University, Budapest

Cernuschi, S and Giugliano, M (1992) Air quality assessments in environmental impact studies in *Environmental Impact Assessment,* Colombo, AG (ed.), Kluwer, Dordrecht

Gabocy, TA and Ross, TJ (1998) Ecological and human health risk assessment: a guideline comparison and review, *Environmental Methods Review: Retooling Impact Assessment for a New Century,* Porter, AL and Fittipaldi, J (eds), The Press Club, Fargo, pp. 193–200

George, C (1997) Assessing global impacts at sector and project levels, *Environmental Impact Assessment Review* **17**: 227–247

George, C (1998) Sustainability assessment through integration of environmental assessment with other forms of appraisal *Impact Assessment in the Development Process* Proceedings of International Conference, University of Manchester, 23–24 October 1998, Manchester, UK

George, C (1999) Testing for sustainable development through environmental assessment *Environmental Impact Assessment Review* **19**: 175–200

Hodson, MJ (1995) Soils and geology, in Morris, P and Therival, R (eds), *Methods of Environmental Impact Assessment*, UCL Press, London

Hope, BK (1995) Ecological risk assessment in a project management context, *Environmental Professional* **17**: 9

Kozova, M, Belcakova, I and Finka, M (1996) Strategic Environmental Assesssment experience in the Slovak Republic, *Strategic Environmental Assessment: Theory versus Practice* Fourth report of NATO/CCMS pilot study, University of Antwerp, Belgium

Landscape Institute/Institute of Environmental Assessment (1995) *Guidelines for Landscape and Visual Impact Assessment*, Chapman and Hall, London

Lumbers, JP (1985) Environmental Impact Analysis in Water Pollution Control, *International Journal of Environmental Studies* **25**: 177

Marriott, BB (1997a) Air quality, in *Environmental Impact Assessment: a Practical Guide*, Marriott, BB (ed.), McGraw Hill, New York

Marriott, BB (1997b) Geology and soils, in *Environmental Impact Assessment: a Practical Guide*, Marriott, BB (ed.), McGraw Hill, New York

Marriott, BB (ed.) (1997c) *Environmental Impact Assessment: a Practical Guide*, McGraw Hill, New York

Ministry of Housing, Spatial Planning and the Environment (1984) *Prediction in Environmental Impact Assessment*, MER Series volume 17 Ministry of Public Housing, Physical Planning and Environmental Affairs, Leidschendam, The Netherlands

Morris, P (1995) Ecology – overview, in *Methods of Environmental Impact Assessment* Morris, P and Therivel, R (eds), UCL Press, London

Morris, P and Therivel, R (eds) (1995) *Methods of Environmental Impact Assessment*, UCL Press, London

National Committee for Integrated Coastal Zone Management (1996) *Integrated Coastal Zone Management in Egypt: towards an Egyptian Framework ICZM programme*, Egyptian Environmental Affairs Agency, Cairo

O'Sullivan, S (1994) Guidelines on assessing the water related aspects of environmental impact statements, *EIS Evaluation Handbook for Local Authorities* Institute of Public Administration, Dublin, pp. 34–43

Ortolano, L (1997) Elements of noise impact assessment, in *Environmental Regulation and Impact Assessment* Ortolano, L (ed.), John Wiley, New York

Patera, A and Riha, J (1996) Environmental impact assessment as a tool for integrated water management *European Water Pollution Control* **6**: 38–49

Radmall, P (1992) *Visual Impact Assessment (VIA): a Review of Current Practice*, The Institute of Environmental Engineering, University of Nottingham, UK

Sadler, B (1996) Environmental Assessment in a changing world: evaluating practice to improve performance, final report of the *International Study of the Effectiveness of Environmental Assessment*, International Association for Impact Assessment, Canadian Environmental Assessment Agency, Quebec

Stamps, A (1997) A paradigm for distinguishing significant from nonsignificant visual impacts: theory, implementation and case histories *Environmental Impact Assessment Review* **17**: 249–293

Tanzania National Parks (1993) *Kilimanjaro National Park General Management Plan*, Tanzania National Parks, Arusha, Tanzania

Tanzania National Parks (1994) *Tarangire National Park Management Zone Plan/ Environmental Impact Assessment*, Tanzania National Parks, Arusha, Tanzania

Therivel (1995) Noise, in *Methods of Environmental Impact Assessment*, Morris, P and Therivel, R (eds), UCL Press London

Treweek, J (1995) Ecological impact assessment *Impact Assessment* **13**: 289–315

United Nations Economic Commission for Europe (1991) *Convention on Environmental Impact Assessment in a Transboundary Context*, UNECE, Geneva

UNEP (1996) *EIA: Issues, Trends and Practice* (prepared by Bisset, R) UNEP, Nairobi

World Bank (1993a) Geographic information systems for environmental assessment and review *Environmental Assessment Sourcebook Update No 3*, World Bank, Washington DC

World Bank (1993b) Sectoral Environmental Assessment *Environmental Assessment Sourcebook Update No 4*, World Bank, Washington DC

World Bank (1994a) Privatisation and environmental assessment: issues and approaches, *Environmental Assessment Sourcebook Update No 6* World Bank, Washington DC

World Bank (1994b) Cultural heritage *Environmental Assessment Sourcebook Update No 8*, World Bank, Washington DC

World Bank (1995a) Coastal zone management and environmental assessment *Environmental Assessment Sourcebook Update No 7*, World Bank, Washington DC

World Bank (1995b) Implementing geographic information systems in environmental assessment *Environmental Assessment Sourcebook Update No 9*, World Bank, Washington DC

World Bank (1996a) International agreements on environment and natural resources: relevance and application to environmental assessment, *Environmental Assessment Sourcebook Update No 10* World Bank, Washington DC

World Bank (1996b) Regional Environmental Assessment *Environmental Assessment Sourcebook Update No 15*, World Bank, Washington DC

World Bank (1997a) Health aspects of environmental assessment *Environmental Assessment Sourcebook Update No 18*, World Bank, Washington DC

World Bank (1997b) Biodiversity and environmental assessment *Environmental Assessment Sourcebook Update No 20*, World Bank, Washington DC

World Bank (1997c) Environmental hazard and risk assessment *Environmental Assessment Sourcebook Update No 21*, World Bank, Washington DC

World Bank (1997d) Environmental assessment of mining projects *Environmental Assessment Sourcebook Update No 22*, World Bank, Washington DC

World Health Organization (1993) *Guidelines for Drinking Water Quality*, World Health Organization, Geneva

6

Economic Valuation of Environmental Impacts

Colin Kirkpatrick

6.1 Introduction

The purpose of Strategic Environmental Assessment (SEA) and Environmental Impact Assessment (EIA) is to ensure that the environmental consequences of development proposals are systematically assessed and taken into account when determining development policies, plans and programmes, and when approving individual projects. The purpose of this chapter is to show how the economic valuation of environmental impacts can be used to support and strengthen the environmental assessment and decision-making process, particularly at the project level.

Earlier chapters have explained how EA tries to identify the effects of development proposals on the environment, and to quantify the magnitude of these impacts, wherever possible. These are measured in a variety of ways, for example particulates per million for air pollution, tons-weight for erosion, decibels for noise pollution, salinization for flooding. But when it comes to overall appraisal at the decision-making stage, these different impacts cannot be directly compared since they are each measured in different units.

The economic valuation of environmental impacts may strengthen the EA process in the following ways:

- Allow the size of different environment impacts to be compared
- Allow different environment impacts to be aggregated into a single measure
- Allow total negative environmental impacts (environmental costs) to be compared with total positive environmental impacts (environmental benefits)

Environmental Assessment in Developing and Transitional Countries. Edited by N. Lee and C. George.
© 2000 John Wiley & Sons, Ltd.

- Provide the basis for clear and defensible decision-making criteria for accepting or rejecting a development proposal
- Promote consistency in environmental assessment and decision-making
- Allow comparisons and rankings of different development proposals in terms of their environmental impact
- Allow environmental impacts to be considered along with the other economic benefits and costs of a development proposal, thereby integrating EA into a more comprehensive cost-benefit analysis (CBA). This is considered further in Chapter 10

The rest of the chapter will:

- Explain the basic concept of economic value (Section 6.2)
- Describe the methods that are available for calculating the economic value of environmental impacts (Section 6.3)
- Explain how intertemporal considerations are taken into account in economic valuations (Section 6.4)
- Provide a summary of the main findings (Section 6.5)

A word of caution should be expressed at this stage. Not everyone agrees with the suggestion that economic values should be placed on environmental impacts. Some object on grounds of principle, arguing that environmental goods and services are inherently different to other goods and services, and are unwilling to base public decisions and choices affecting the environment on economic values. Others accept that it would be useful to know the economic value of environmental goods and services, but do not believe that existing knowledge and techniques are sufficiently well developed to measure these economic values accurately.

This chapter takes an intermediate position, arguing that some but not all environmental impacts can, and should, be measured in economic terms. There are various methodological and practical difficulties associated with the economic valuation of the environment. Nevertheless, economic valuation, provided it is carried out carefully and applied sensibly, can improve the transparency, consistency and overall quality of decision-making. There will still be environmental effects which are non-quantifiable, and there will continue to be a need for judgements based on quantifiable and non-quantifiable information, perhaps using multi-criteria analysis (see Chapter 10). However, the difficulty of making these judgements will be reduced the more environmental impacts can be expressed in economic terms (Winpenny 1991, pp. 6–7).

6.2 The Concept of Economic Value

Economics is about making choices. Choices have to be made when resources are limited and it is scarcity, therefore, that imparts economic value to a good or service. In the market place individuals make choices on how to use their limited resources (i.e. their income and wealth) by comparing the price of a good or service with their subjective judgement of how much the acquisition and consumption of the item will contribute to their welfare or well-being. If an individual is willing to pay a given amount of income for a good, then it may be assumed that the willingness to pay is equal to the value which s/he attaches to the good or service in question. *Willingness to pay*, therefore, measures his/her economic value.

Thus, the economic value of an environmental impact to an individual is the money payment that s/he is *willing to pay* (WTP) in order to secure it (if positive) or to avoid it (if negative). Alternatively, it is the money payment s/he is *willing to accept* (WTA) as compensation for an environmental impact (if negative).

Most goods and services are bought and sold by individual consumers and producers at a market price. The market price can then be used to calculate the individual's willingness to pay and hence, the economic value of the good to that person. The same principle underlies the economic valuation of environmental goods and services.

However, many environmental assets are not traded in a market, and therefore do not have a market price. Important environmental goods, such as air, oceans, biodiversity, commonland, are unpriced because they are *public goods*, which are available to everyone and which cannot readily be denied to anyone. As a consequence, it is impossible to charge for their use. Although environmental public goods are available to all, they are also often depletable – one person's use reduces the availability to others. In other words, there is a scarcity problem, choices have to be made, and the environmental good acquires an economic value. In the absence of a market price, the task of the environmental economist is to find alternative methods of estimating consumers' willingness to pay for environmental assets and the services they provide.

Environmental public goods are just one example of *market failure*, where the market mechanism fails to signal the relative scarcity and economic value of different resources through their prices. There are other causes of market failure which can also affect the economic valuation of environmental assets. These are summarized in Box 6.1.

Willingness to pay (and therefore, economic value) is also influenced by *ability to pay*. For this reason, low income countries typically place lower economic values on environmental impacts than do high income countries. Similarly, poorer communities place lower values on environmental impacts than do richer communities within the same country. Some economists recommend, on equity grounds, that economic values should be adjusted to take account of income distribution considerations. Others suggest that it would be more appropriate to address the causes of the income inadequacies more directly rather than the consequences which flow from these. An intermediate position is to retain market-based economic values but to present the findings in a disaggregated form so that their distributional consequences are clear to decision-makers (see Chapter 10 for a further discussion of the handling of distributional issues in appraisal and decision-making).

Components of Total Economic Value

An alternative approach to calculating the total economic value (TEV) of the environmental impacts of a project is to disaggregate the impacts into individual components

---

Box 6.1   Market Failure and the Environment

Market failure can be complete, as in the case of public goods. Market failure also occurs when markets generate prices that do not sufficiently reflect the full economic value of goods and services. Such prices convey misleading information about resource scarcity and provide inappropriate incentives for the management, efficient use, and conservation of environmental resources.

Sources of Market Failure

- *Public goods.* A good that is available to everyone and as a consequence it is impossible to charge for its use. The use of the good however, often results in its depletion. Example: fish caught in international waters

- *Externalities.* The effects of a development on other parties that are not taken into account by its owner. Example: a factory discharges harmful effluent into a river used for drinking and fishing by the local community

- *Ignorance and uncertainty.* Environmental processes are often poorly understood, and markets may fail to signal emerging resource scarcities and future environmental consequences. Example: the impact of increasing atmospheric pollution on climate change

- *Short-sightedness.* Consumers and producers may not be interested in long-term environmental costs and benefits which are likely to occur after their lifetime. Example: neglect of long term costs of de-commissioning nuclear power facilities

- *Irreversibility.* Certain environmental changes cannot be reversed. It may be desirable to maintain the option of using a non-renewable or replaceable asset at a later date but markets will not readily allow for this option value. Example: the flooding of a scenic valley for a hydro-electric scheme which effectively removes the option of preserving the existing landscape for future generations

Source: OECD 1995.

---

of value. The reasoning behind this approach is that environmental impacts have different attributes, some of which can be more easily measured than others.

Figure 6.1 distinguishes between five components of total economic value. In general, as one moves from left to right in the diagram, the difficulty in estimating economic value increases, with direct use value being the most straightforward component of TEV to estimate and existence and bequest values being the most difficult.

Direct use value

Direct use value refers to benefits that accrue from direct use of the environmental asset in question. This is generally the easiest component of TEV to value, where it relates to observable quantities of resources or products whose market prices can be observed.

Indirect use value

Indirect use value derives from the services that the environment provides, in addition to and separately from, direct use value. For example, a wetland may act as

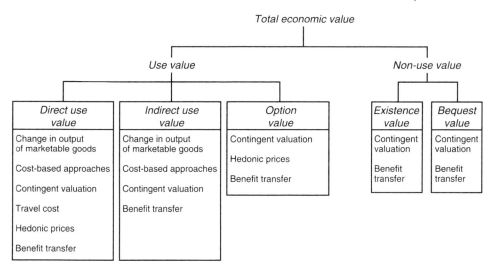

Figure 6.1 *Total Economic Value and Selected Valuation Techniques.* Source: World Bank (1998)

a filter, which indirectly improves water quality for users, or a forest may indirectly help to protect watersheds.

Option value

Option value is the value obtained from retaining an option on the future use of an asset. No use is made of the asset now but there is value in retaining it for possible use in the future.

Existence and bequest value

Existence and bequest values are both non-use value elements in TEV. The non-use value derives from the benefits that the environment provides which do not involve using it in any way. Existence value is derived from the knowledge that something exists. Thus, people place a value on the protection of rare animals from extinction, even if they do not expect to see them. Bequest value is the value derived from the knowledge that something is being passed on to one's descendants.

The example of a tropical forest can be used to illustrate these different types of economic value. The forest has direct use value in the form of timber and pharmaceutical products, and recreational uses. It also has indirect use values, including flood and storm protection, and air quality protection through the absorption of carbon dioxide. The option value is provided by the retention of the forest for potential future use. Existence and bequest value is derived from the knowledge of the forest's continued existence and its intrinsic value for cultural and other purposes to present and future generations.

6.3   Valuation Techniques

There is a range of valuation techniques that can be used to estimate the five components of total economic value shown in Figure 6.1. The choice of technique will depend on the particular environmental impact under consideration and on the availability of data. In some instances, it may be possible to apply several techniques to the valuation of the same environmental impact, which can provide a useful cross-check on the reliability of the estimates obtained.

In all cases, the underlying approach is the same – to estimate what individuals would be willing to pay (or willing to accept in compensation) for a specified change in an environmental good or service. There are three main ways of calculating these values:

1. Using market prices.
2. Using information on individuals' preferences.
3. Benefit transfer.

Valuation Using Market Prices

Change in productivity

This method values environmental change by observing physical changes in the environment and estimating what difference they will make to the value of marketed goods and services. This approach is applicable in calculating direct and indirect use value. Water pollution can reduce fish catches, and air pollution can affect the growth of crops. In both instances, the environmental impact reduces marketed output, which may be valued using market prices.

Box 6.2   Coastal Forest Protection Project, Croatia

Reforestation activities were estimated to result in increased wood production, which would be harvested at various intervals in the future. Using estimates of increased output (in terms of quantity and quality) and expected price at the time of harvest, it was possible to calculate the economic value of the increased wood production.

Source: World Bank 1998.

Human capital cost valuation

This method may be used to value the impact of environmental hazards on human health. Environmental 'bads' such as air and water pollution or the use of pesticides reduce the quality of the human capital stock, and therefore lower the economy's productive capacity. To apply the human capital cost method it is first necessary to determine the relation between the hazard and human health, by expressing the health impact in terms of premature death, sickness or absenteeism. Sickness can then be valued using medical and health care costs. Absenteeism is valued in terms of lost earnings (this assumes that earnings measure the contribution that the absent worker would have made to output).

Box 6.3   Air Pollution Costs in Mexico City

A 1991 World Bank study used the cost-of-illness approach to estimate air pollution costs. The study used a three step procedure:

- Determining the ambient concentrations of various pollutants
- Using dose–response relationships to determine the incremental incidence of disease including both morbidity and mortality in the population
- Estimating the costs of the increase in morbidity and mortality, as measured by treatment costs, loss of wages and loss of lifetime earnings

Source: Dixon *et al.* 1994.

Valuation Using Information on Individuals' Preferences

Often it will not be possible to link the environmental impact to a change in marketable output. In these cases, the willingness to pay has to be estimated indirectly, using a range of other techniques, such as:

Replacement cost or preventive expenditure method

The economic value that individuals attach to the environment can sometimes be inferred from the cost of preventing unwanted environmental impacts, or of restoring an asset to its original state after it has been damaged. For example, the costs of air pollution-related acid depositions could be estimated using the costs of restoring damaged physical infrastructure, or the costs of soil erosion could be estimated using the costs of providing preventive terracing.

Box 6.4   Flood Control and Soil Conservation Project, Yellow River Basin, China

This was intended to reduce flooding and deposition of sediment in the lower reaches of the Yellow River by a number of measures undertaken in the upstream area: construction of structures to trap sediment; modification of land form; modification of land use.

The flood prevention benefits were valued indirectly in terms of avoided expenditures – preventive expenditure on raising dikes, restoration costs of desilting irrigation systems, and the opportunity cost of water used for flushing sediment.

Source: Dixon *et al.* 1994.

Contingent valuation method

The contingent valuation method (CVM) relies on direct questioning of people to determine their willingness-to-pay valuation of an environmental impact. A detailed description of the environmental impact is provided, and interviewees are then asked what they would be willing to pay (WTP) for a hypothetical environmental improvement, or to accept (WTA) as compensation for an environmental deterioration.

The contingent valuation approach may, in principle at least, capture the total economic value (use and non-use components), whereas other techniques may only provide estimates of direct or indirect use value. However, CVM also has a number of shortcomings (see Table 6.1).

Box 6.5   National Park Project, Madagascar

CVM was used to value the loss of benefit to local communities of refraining from using the Mantadia National Park. Local residents were asked whether they would be willing to accept specified levels of compensation to forego access to the forest. These estimates were then used as a measure of the costs imposed on the local community by the loss of access to the Park. CVM was also used to estimate the benefits to international tourists from visiting the Park.

Source: World Bank 1998.

Surrogate market valuation method

Whilst an environmental good or service may not be traded directly, it is sometimes possible to find a good or service, related to the non-marketed environmental item, that is sold in markets. In this situation, the individual may reveal his or her preference for both the market and non-market good or service when making a purchase. It may then be possible to separate-out the environmental component of value from the observed market price, and in this way use this component of market price as a 'surrogate' for the environmental value.

There are two main techniques which have been used for applying the surrogate market method: travel cost method and property value (hedonic price) method. Each method is described, together with examples of their application in developing countries.

Travel cost method. Many natural resources (e.g. a national park or lake) are used for recreational purposes. The travel cost method bases its valuation on the money and time costs of visitors to such recreational facilities.

Box 6.6   Elephant Viewing Safaris in Kenya

The travel cost method was used to estimate the value of safaris and the contribution that elephants make to this value. The cost of travel was estimated using data on land travel costs, air fares and travel time. To identify the contribution that elephants make to the value of a safari, tourists were asked to allocate the enjoyment of their trip over various categories of experience, including viewing elephants. The proportion attributed to elephant viewing was applied to the total travel cost valuation to give a viewing value for elephants.

Source: Munasinghe 1993.

Table 6.1   *Environmental Valuation Methods*

Valuation method	Advantages	Disadvantages
Change in output of marketable resources and goods	• Easily understood and applicable, provided dose-response relation is known • Uses actual market prices	• Difficult to isolate the effect of given impact on observed change in production • Market prices may be a poor indicator of willingness to pay • Only relates to use value
Human capital cost	Applicable where: • Epidemiological dose-response data • Health expenditure data, and • Earnings data are available	• Likely to understate full value of health • Difficult to isolate separate causal factors in ill health • Moral and ethical objections
Cost based approaches (Replacement cost or preventive expenditure)	• Ease of application, if relevant technical and cost data are available	• Preventive expenditure may understate environmental value • Replacement cost may understate full reinstatement of environment quality • May not cover non-use values
Contingent valuation	• Potentially covers most components of total economic value • Practice improving with greater experience in its use	• Time-intensive and expensive to implement • Biases through use of stated rather than revealed preferences • Other biases associated with questionnaire design and survey practices
Travel cost	• A fairly well developed and used method	• Significant data requirements • Problems in reliably interpreting the statistical findings • Measures use value only
Property valuation/ hedonic pricing	Applicable where there is: • Availability of property price data • Availability of data relating to determinants of property prices	• Assumes market values capture the environmental good's value • Problems in segregating the influence on property prices of environmental factors from that of other explanatory variables • Measures use value only
Benefits transfer	• Time saved and inexpensive • Applicable where value estimates are available from other comparable studies	• Inappropriate transfer of values from sites where primary analyses were conducted to sites experiencing different, non-comparable conditions

Property value (or hedonic price) method. The hedonic price method is based on the idea that differences in property prices can be used to infer the value which individuals attach to the difference in environmental quality between properties. For example, the difference in the price of two properties which differ only in, say, the local air quality, will provide a measure of the value which people give to difference in air quality. Even when properties differ in other ways, it may still be possible (though it is a complex task) to uncover the implicit prices of environmental quality using statistical techniques to separate out the contribution of each factor to the total market price.

Box 6.7   Slum Improvement Project, Visakhapatnam, India

In 1988 the UK Overseas Development Administration started a major programme to improve 170 designated slum areas. The programme included physical infrastructure improvements, improved water supply, public toilets, community centres and primary health care services. The average change in property prices over a three year period was calculated for slum areas that were included in the improvement programme, and for areas that were not included. The difference in values was taken as a measure of the benefits accruing from the slum improvement scheme.

Source: OECD 1995.

Benefit Transfer

Benefit transfer involves deriving estimates of economic value in one context for use in a different context, where the data required for the estimation are not readily available. For example, the value of health damage from air pollution in one city might be used to estimate health costs from air pollution in a different city or, more controversially, the values derived in one country might be transferred for use in a different country. Though this can provide quick and low-cost estimates, it is subject to a number of limitations (see Table 6.1)

Table 6.1 summarizes the main valuation techniques and lists some of the advantages and disadvantages of each method.

6.4   Intertemporal Considerations: Choice of Discount Rate

Economic values that arise in the future may have a lower value than the equivalent value arising in the present. Discounting is the process whereby future values are converted into their equivalent present-period values, using discount rates, which then allows the stream of future values (benefits and costs) to be aggregated into a single net present value sum (NPV) estimate.

Some environmentalists suggest that the discount rate applied to environmental projects should be lower than that used for other economic projects. It is argued, for example, that significant environmental benefits and costs are likely to occur in the long term, and would be reduced to insignificance in present-value terms if a conventional discount rate is applied. It is also argued that the use of a conventional discount rate would encourage the short-term exploitation of renewable natural resources, such as forests and fisheries. Therefore, a lower discount rate should be applied to environmental effects and projects to counteract the 'anti-future' bias of conventional discounting and to safeguard the environmental interests of future generations.

Others oppose the use of a lower discount rate for environmental effects. First, they argue, the overall pattern of resource use and investment choice will become distorted if the environmental impacts of projects are treated differently from other types of impacts. Second, a low rate of discount may allow additional projects to be undertaken, which may increase the overall pressure on the environment. Finally, the increasing scarcity of resources can be counteracted more appropriately by increasing their bequest and existence values.

The prevailing opinion, among economists at least, is that environmental concerns are better handled by the appropriate valuation of the environmental effects rather than by discriminatory adjustments to the discount rate. But this may need to be qualified after giving due consideration to sustainability criteria. Using irreplaceable natural capital should be avoided as far as possible; irreversible processes of environmental deterioration should also be avoided in order to keep options open for future generations. The application of sustainability criteria may mean for example, preserving biodiversity, avoiding extinction of species, and maintaining valued landscapes and habitats. In cases such as these, decision-making may need to be based on sustainability criteria, rather than on discounted net present values alone.

6.5 Conclusions

This chapter has provided a rationale for estimating the economic value of environmental impacts at the project level. The principal valuation methods have been described, their strengths and weaknesses identified, and their applications illustrated.

However, it has to be acknowledged that the economic valuation of certain environmental impacts is difficult to undertake in practice and that the valuation findings which are obtained may not be widely acceptable. Certain of the valuation techniques which are used may assume conditions that are not easily met, or may require data that are not readily available, in many low and middle income countries. Also, the handling of equity considerations within the valuation process may be problematic in these countries.

In principle, economic valuation methods can also be applied in the strategic environmental assessment of policies, plans and programmes. However, additional complexities to those which arise in project-level assessment need to be taken into consideration:

- The predictions of the environmental impacts which are to be valued are subject to greater margins of error at this level. This compounds the uncertainties associated with the use of the economic valuations themselves (see Chapter 5, this volume; Lee and Kirkpatrick 1999)

- Strategic-level PPPs are more likely to change relative prices and complicate the choice of appropriate market prices to use for valuation purposes (Persson and Munasinghe 1995; Chowdhury and Kirkpatrick 1994, Chapter 6).
- The distributional consequences of certain PPPs (e.g. structural adjustment programmes) could be extensive and, on equity grounds, may need fuller consideration in the economic valuation of their environmental consequences (Reed 1996).

Thus, the analyst may find that economic valuation can only be satisfactorily applied to certain of the environmental impacts associated with a particular development proposal. The remainder will have to be quantified (or, failing this, expressed qualitatively) in non-economic terms. Within these limits, the careful application of economic valuation techniques can enhance the information on, and understanding of, the significance of the environmental consequences of a development proposal. However, the analyst will also need to use additional methods, as well as good judgement and common sense, when assessing whether to proceed with a given development project or policy proposal. These additional methods and approaches are discussed, inter alia, in Chapters 7, 9 and 10.

Discussion Questions

1. Why are some economic values reflected in the market place and some not?
2. What is meant by the 'willingness to pay' measure of economic value? How can 'willingness to pay' for environmental benefits be estimated?
3. Should environmental benefits and costs be discounted at different rates than other economic benefits and costs? Should environmental impacts on low income communities be valued differently than impacts on high income communities?

Further Reading

Introductions to the basic principles of economic valuation of environmental impacts in a developing country context are found in OECD (1995) and World Bank (1998). A more advanced treatment is provided in Georgiou *et al.* (1997) which also contains a lengthy annotated bibliography. A useful Manual/Workbook on the application of economic valuation techniques in EA is found in ADB (1996). Case studies of the application of economic valuation in developing country projects are found in Dixon *et al.* (1994), Winpenny (1991) and Munasinghe (1993). Kirkpatrick and Lee (1997) and Lee and Kirkpatrick (1999) focus on a range of issues relating to the integration of environmental assessment and socio-economic appraisal in developing countries and countries in transition.

References

Abaza, H (1997) Integration of sustainability objectives in structural adjustment programmes through the use of strategic environmental assessment, in *Sustainable Development in a*

Developing World: Integrating Socio-Economic Appraisal and Environmental Assessment, Kirkpatrick, C and Lee, N (eds), Edward Elgar, Cheltenham

Asian Development Bank (ADB) (1996) *Economic Evaluation of Environmental Impacts: A Workbook*, ADB: Manila

Chowdhury, A and Kirkpatrick, C (1994) *Development Policy and Planning: An Introduction to Models and Techniques*, Routledge, London

Dixon, J, Scura, LF, Carpenter, RA and Sherman, PB (1994) *Economic Analysis of Environmental Impacts*, 2nd edn, Earthscan, London

Georgiou, S, Whittington, D, Pearce, D and Moran, D (1997) *Economic Values and the Environment in the Developing World*, Edward Elgar: Cheltenham

Kirkpatrick, C and Lee, N (eds) (1997) *Sustainable Development in a Developing World: Integrating Socio-Economic Appraisal and Environmental Assessment*, Edward Elgar, Cheltenham

Lee, N and Kirkpatrick, C (eds) (1999) *Sustainable Development and Integrated Appraisal in a Developing World*, Edward Elgar, Cheltenham (in press)

Munasinghe, M (1993) Environmental issues and economic decisions in developing countries *World Development* **21**: 1729–1748

Munasinghe, M and Cruz, W (1995) Economy-wide policies and the environment: lessons from experience *World Bank Environment Paper* no. 10, World Bank, Washington DC

OECD (1995) *The Economic Appraisal of Environmental Projects and Policies: A Practical Guide*, OECD, Paris

Persson, A and Munasinghe, M (1995) Natural resource management and economy-wide policies in Costa Rica: a computable general equilibrium (CGE) modeling approach, *World Bank Economic Review* **9**: 259–286

Reed, D (ed.) (1996) *Structural Adjustment, the Environment and Sustainable Development* Earthscan Publications, London

Winpenny, J (1991) *Values for the Environment: A Guide to Economic Appraisal* HMSO, London

World Bank (1998) Economic analysis and environmental assessment, *Environmental Assessment Sourcebook Update* no. 23, April, World Bank, Washington DC

7

Social Impact Assessment

Frank Vanclay

7.1 Introduction

Social Impact Assessment (SIA), sometimes called social assessment, is well established in many countries. However, the issues, concerns and contexts of projects in developing countries are often different from those in more developed countries. Also, developing countries themselves are not homogenous (between and within countries). In this short chapter, an attempt is made to address issues relating to SIA in developing countries, but without elaborating on SIA in general due to the availability of much commentary elsewhere on this subject (e.g. Burdge 1994; Burdge and Vanclay 1995; Taylor *et al.* 1995; Vanclay 1999).

SIA is a broad area of knowledge that contains different understandings. It is usually defined according to the purpose to which it is being put, by whom it is being used and in what context. The nature of an SIA undertaken on behalf of a multinational corporation, where it is a requirement of that company's internal procedures, may be very different to an SIA undertaken by development agency officials interested in 'best value' for their country's development assistance. This, in turn, may be very different to an SIA undertaken by staff or students at a local university on behalf of a local community, or an SIA undertaken by the local community itself. Each of these uses of SIA has value, but none should be regarded as the sole definitive statement. Indeed, whenever an SIA is being evaluated, its intended purpose or role should always be taken into consideration.

In western countries, SIA tends to be defined along the lines of 'the process of assessing or estimating, in advance, the social consequences that are likely to follow from specific policy actions or project development, particularly in the context of appropriate national, state or provincial environmental policy legislation' (Burdge

Environmental Assessment in Developing and Transitional Countries. Edited by N. Lee and C. George.
© 2000 John Wiley & Sons, Ltd.

and Vanclay 1995, p. 32). However, this is not the most useful understanding of SIA when considering projects in developing countries. Here, SIA needs to be considered more as 'a framework for incorporating participation and social analysis into the design and delivery of development projects' (World Bank 1995a) or as 'a process for research, planning and management of change arising from policies and projects' (Taylor *et al.* 1995, p. 1). Thus, SIA needs to be process oriented, and to ensure that social issues are included in project design, planning and implementation as well as ensuring that development is acceptable, equitable and sustainable (Branch and Ross 1997). The improvement of social wellbeing, with a particular focus on poverty reduction and an emphasis on democratisation, should be explicitly recognized as an objective of development projects and plans, and as such should be a performance indicator considered in any form of impact assessment.

A first purpose of SIA is the anticipation of potential negative impacts of various development options to allow for an informed choice of the best option for society as a whole. A second purpose is to maximise benefits and minimize negative impacts by modification to plans (mitigation), partly through the use of local knowledge. This is best achieved by a participatory process which may also assist in the reduction of opposition to the project (and thereby reduce costs to the developer). SIA also enables an assessment to be made of the distribution of costs and benefits, leading to the modification of projects, to ensure that the same people are not always negatively impacted. Development often involves the transfer of technology between cultural settings. The SIA process should facilitate understanding of the local culture with a view to ensuring that there is no fundamental mismatch between development and the local culture. Where there is, it should advise on changes to that technology or on the implementation of a change management plan.

7.2 The Methodology of SIA

There is no universal, methodological standard for SIA currently available (possibly Interorganisational Committee (1994) comes closest to this for developed countries). However, in many contexts, a good SIA is likely to include the following:

- Identifies interested and affected peoples (IAPs or stakeholders)
- Facilitates and co-ordinates the participation of IAPs
- Documents and analyses the local historical setting in which the project will occur
- Provides a rich picture of the local cultural context and an understanding of local community values
- Identifies and describes the activities which are likely to cause impacts (scoping)
- Predicts likely impacts, including cumulative impacts, and how the community might respond
- Assists in the selection and evaluation of programme alternatives (including a no development option)
- Assists in site selection
- Recommends mitigation measures

- Provides suggestions relating to compensation
- Describes potential conflicts between stakeholders and advises on conflict resolution processes
- Develops strategies in the community for dealing with residual or non-mitigatable impacts
- Contributes to skill development and capacity building in the community
- Advises on appropriate institutional and co-ordination arrangements for all parties
- Assists in the devising and implementation of monitoring and management programmes
- Collects data for profiling to allow evaluation and audit of the impact assessment process and the project itself

7.3   Role of Participation in SIA

Participation and other forms of public involvement are not synonymous with SIA but they are an important constituent of it. Participatory approaches have four main advantages:

- They provide a better understanding of local values, knowledge and experience (a common error among western impact assessors is to assume that local people, in different cultures and socio-economic conditions, have the same values as they do)
- They help create community support for the project
- They help resolve conflicts
- They help the community to understand the project and its implications, to plan for the change, and adapt to the changed situation

Participation does present some difficulties. In many countries there is not a culture of participation, either because it is not part of the social culture, and sometimes because it has not been part of the political culture. In transition states, and in some regions which have had or currently have repressive regimes, promoting meaningful participation may be difficult. No matter how good the intentions of the practitioner, it may be hard to convince local people that there will not be adverse consequences for them as a result of their involvement.

In some countries, difficulties include the lack of familiarity of the affected peoples with the proposed project. For example, what do high voltage transmission corridors mean to people who do not know what electricity is? What would a nuclear reactor or nuclear weapons testing facility mean to them? However, there are ways of conveying some impression of the project and the magnitude of the impacts that might be anticipated from it. Such participation processes do consume extra time, and may require creative thinking. In all cases, the form of participation must be appropriate to local cultural traditions.

While evaluations of projects in many continents reveal that successful projects are those in which there has been considerable participation (Bass *et al.* 1995), participation is not a universal panacea and cannot guarantee success. There may be unresolvable conflicts due to the way in which a project divides a community into winners and losers, or because of background tensions that preceded the current project. In order to be sensitive to this type of situation, an understanding of the local history is essential.

It is also important to be aware of the possible misuse of the 'participation' provisions contained in competent authority or funding agency guidelines. Unfortunately, the reality of many development activities is one of limited or token consultation in which participation cannot materially influence the outcomes of the project (Peiris 1997). Such practices could have repercussions on the acceptability of other current and future projects as local people become disillusioned with the process.

7.4   Why a Gender Analysis is Essential

Household units are often used as the basis of analysis but ultimately all impacts are experienced by individuals. People experience impacts in different ways depending on their social situation. One important facet of this is the gendered nature of life experiences. Unfortunately, gender-blindness is a condition that has afflicted many development projects and impact assessments (Guijt and Shah 1998). However, it is also vital to appreciate the diversity in situation and experience among different types of women, as well as different types of men.

Although gender is not the only social variable explaining the differential experience of impacts, it is one of the most important variables. When it is predicted that the level of work required to survive will increase because of changes to the environment brought about by a particular project, it is often the women who will bear the brunt of increased loads. If men leave their homes and villages for work at a distant workplace, it is the women who remain who face an increased workload. When men earn extra money, cash is often expended on consumer goods (e.g. televisions), rather than leading to a reduction in women's workload (O'Rourke 1980). If entrepreneurial activities are promoted to allow villagers to earn additional cash income, it is often the women who experience an increased workload to earn this income. Thus, it can be argued that many development projects have worsened the position of women (Nabane 1995).

Men and women tend to have different roles, with women being responsible for many of the farming activities, tending animals, firewood collection, water collection and cooking. Environmental impacts have serious social impacts when they reduce food, water and/or firewood, because they dramatically increase the workload on women (Banuri and Holmberg 1992). Some of the biggest impacts relate to increasing the distance to collect firewood or to collect fresh water. At the same time, women's daily collection activities traditionally were social occasions, when they were able to mix with other women. Supplying water and gas directly into people's houses removes these social occasions and may isolate women in their houses. Public wells or water points close by villages may be more appropriate. However, with increasing western influence in many countries, local aspirations may be changing, causing conflict between social interaction and personal aspiration, and creating a dilemma for social planners.

When countries implement macro-economic readjustments and reduce public expenditure programmes, amongst the first casualties are the social services that assist women and children. Women, being poorer, are usually more affected by declining economic standing. In the employment marketplace, women are less skilled (at least in terms of skills with market value) and their other skills are undervalued. Consequently women not only have lower pay but also lower security of employment and are usually the first to be dismissed in any economic restructuring process (Moon 1995).

A final reason for a gender analysis is that because of different duties, and women's connection to the land, especially in hunting and gathering societies, women's local ecological knowledge may be greater than that of men (Nabane 1995). Consequently, the role of women in SIA teams needs to be considered, and the way in which participation is structured needs to ensure opportunities for the contribution of women. There are often local cultural rules which govern who is allowed to talk to whom, especially in the presence of others. Generally, women should be included in all SIA teams to discuss issues relating to development impacts with other women.

7.5 'Content' of SIA in Developing Countries

A simple way of describing the nature of social impacts would be as changes in one or more of the following (Vanclay 1999):

- People's way of life – how they live, work, play and interact with one another on a day-to-day basis
- Their culture – shared beliefs, customs and values
- Their community – its cohesion, stability, character, services and facilities
- Their environment – the quality of the air and water people use, the availability and quality of the food they eat, the level of dust and noise to which they are exposed, the adequacy of sanitation, their safety and fears about their security, and most importantly, their access to and control over resources

Thinking about these broad areas of impacts leads to an appreciation that all major types of social impacts are potentially included. There have been many attempts to develop detailed lists of social impacts (Burdge 1994; Vanclay 1999). However, there is a concern that the process of identifying impacts should not degenerate into the naïve, mechanical use of checklists. Scoping processes, with local community representation, should be carefully and intelligently used to determine which are the important impacts in any particular situation.

7.6 Some Important Considerations in Cross-cultural Assessments

Many tropical countries have extremes of seasons (wet and dry) such that the impacts that are important in one season may be very different to those that are important in another season (Chambers 1983, 1993). The severe humidity of the wet season may cause machinery to malfunction, and torrential rain may prevent access. The wet season is also a time when the ecology of disease organisms and vectors is altered which may radically alter the patterns of disease in a community. The dry season may be a time of hardship because of the lack of water and fodder for stock. If drought persists, food and water reserves for humans will also be reduced.

Many development initiatives have attempted to raise the productivity of food production systems. However, many of these have failed because of insufficient appreciation of local ecological and social systems. For example, attempts to convert subsistence agriculture to cash cropping systems have exposed farmers to the vagaries of world market fluctuations and have led to greater reliance on monocultural production regimes, with attendant insect and weed problems as well as land degradation. While cash crops have provided a supply of income and foreign currency, they have often led to a reduction in the locally available supply of food. Part of the income earned has been spent on imported consumer goods, which has added to the foreign exchange problem that the cash crops were meant to alleviate. Cash crops, and new varieties of certain staple crops, tend to rely more heavily on imported agricultural inputs such as pesticides, herbicides and fertilizers. Other impacts may include the loss of local skills and crafts, increased inequalities in income and wealth, and the disintegration of local culture through the adoption of western materialistic values. Increased financial pressures on the poorer members of local societies have led to greater dependence on village money-lenders charging high interest rates.

In western countries, land tenure systems are usually clearly defined by the legal system and seldom challenged, unlike in many developing countries where different groups of people will claim ownership of, or control over, land and other resources. Even within groups, there may be disputes between individuals over which individual is the rightful owner. Often, land ownership (or control over access to land) has, by tradition, been communal. Western notions of ownership have been imported into these cultures where they were previously unknown and they cannot easily resolve such issues as compensation for loss of land due to development.

Approval from one authority structure doesn't necessarily imply approval by all authorities. Traditional authority structures are often patriarchal and religious in nature. Participation strategies are needed to negotiate with these authority structures, but also to consult beyond these structures in order to ensure that the concerns of disenfranchised peoples are also considered.

Some developing countries have cultures that abhor conflict, explicit criticism or even the questioning of authority. Etiquette may demand that people always indicate agreement with an outsider even in situations where there may be profound disagreement. These issues limit the validity of many public participation processes, especially if implemented by an outsider. Both qualitative and quantitative methods of inquiry are potentially affected by the cultural reluctance of people to openly criticize superiors.

In most developing countries, land is designated as belonging to one or more groups, even if it appears comparatively empty. Development schemes that require resettlement of peoples, or that lead to the migration of people, are fraught with problems because of inevitable conflict between the people resettled and the traditional owners of the land. Because economic systems in developing countries are not exclusively cash economies, financial compensation is often not feasible, at least not in the way in which it is conceived elsewhere. In any case, compensation schemes have their own social impacts and should not replace attempts at mitigation (Vanclay 1999).

Murder, 'disappearance', torture and other forms of intimidation are evident in some countries. A culture of suspicion may exist, brought about by the presence (or fear) of government spies and concern over possible government reprisals. In such circumstances, participation may be difficult, and attempts to seek participation may create unease. There may be considerable corruption and significant proportions of government finance for development may be misappropriated. However, where the offering of gifts is part of the legitimate cultural process, care is needed to differentiate this from the corrupt use of funds.

One of the concerns with development projects relates to their appropriateness for the societies which they should serve. Many energy development projects, for example, are for the production of electricity, but this may be used for such purposes as air-conditioning, which only benefits more affluent urban minorities. The increasing urbanization of many cities in developing countries, has been accompanied by an increased consumption of resources, often at the expense of the rural poor (Fuggle 1995).

The poorest people are often fringe-dwellers. They may not be legal residents, and sometimes may not even be citizens of the country in which they reside. They include those who have migrated in search of work but also dispossessed people and refugees escaping hardships (due to war or famine) in their own homeland. The distributional consequences, for these poorer sections of society, of proposed developments also need to be investigated as part of the SIA process.

An important component of the development strategy of many countries is the expansion of tourism. However, this can have considerable social impacts due to cultural conflicts between tourists and the local peoples, as well as to increased resource use and waste generation associated with tourist activities (Cohen 1983; Cohen 1984; de Kadt 1979; Mathieson and Wall 1982; Phillip 1983; McLaren 1994; Edensor 1998; Munt and Mowforth 1998; Wahab and Pigram 1998). Among these cultural impacts are those arising from the marketing of artefacts, and the sale of sacred objects to tourists. Major inequalities in income and wealth, between tourists and local people, can lead to considerable distortions in the local economy with unforeseen social consequences, as well as to local resentment. Increased tourism can also lead to a rise in prostitution (Truong 1990) with its attendant social and health concerns, as well as increased criminal activity. While eco-tourism can lessen negative impacts, it also has some negative social and environmental consequences of its own (Roe *et al.* 1997).

It is often suggested that there are insurmountable difficulties in undertaking good quality SIA/EIA in developing countries. For example, the problems raised are that: a large percentage of the population may be illiterate, many may be struggling to fulfil basic needs, there may be conflict between a nation's goals of economic development and environmental protection, EA procedures may be seen as too expensive, there may be a belief that appropriate data do not exist, and that expertise is not readily available. MacDonald (1994) argues that many of these difficulties can be overcome through the use of participatory strategies. However, others argue that the difficulties are more ideological in nature and stem from an asocietal attitude among policy makers, development agencies and EIA practitioners (Burdge and Vanclay 1995).

7.7   Conclusions

> Putting people first in development is not just a slogan. It is a shorthand for a different development philosophy, and for an operational approach that means starting with people and building projects and programs around their needs and capacities to act. (Cernea 1993, p. 9)

SIA and EIA practitioners need to be reminded of the importance of this. In particular, they need to examine critically 'the econocratic and technocratic models that still inspire – and distort – development interventions in many agencies and government offices' (Cernea 1993, p. 2). By 'econocratic' models, Cernea (1993, pp. 2–3) means 'interventions that focus one-sidedly on influencing the economic variables, regarding them as the only decisive ones and assuming that the "rest" will necessarily "fall into place"'. By 'technocratic' models, he means the 'projects that address the technological variables more or less "in-vitro", dis-embedded and dis-embodied from their societal context'. In their place, there is a need to establish a development paradigm that is centred on people. This will also prove challenging to some environmental professionals who consider that ecological issues are of great importance and override all other considerations.

A full consideration of social issues and social impacts is very challenging. Participation practices are not easy, and they are time consuming. Nor can they guarantee satisfactory results. But they are essential to the long-term realization of sustainable development.

Discussion Questions

1. What are the main differences between the types of development impacts experienced by people in western countries and those experienced by people in developing countries?
2. What are the main reasons why people in different societies, developed and developing, value impacts differently from each other?
3. Why is a gender analysis a necessary component of a satisfactory SIA?

Further Reading

References on SIA in general include Burdge and Vanclay (1995), Taylor *et al.* (1995) and Vanclay (1999). There are few publications that relate specifically to

SIA in developing countries but some worthwhile references include Cernea (1991), Cernea and Guggenheim (1993), Cernea and Kudat (1997), World Bank (1995b) and ODA (1995). Useful references relating to participation in development are Bass *et al.* (1995), Burkey (1993), Engel and Salomon (1997), Holland and Blackburn (1998), Pretty *et al.* (1995) and Salmen (1987). Useful works that relate to gender analysis include Feldstein and Jiggins (1994), Guijt and Shah (1998), Slocum *et al.* (1995) and Royal Tropical Institute (1998). Shiva (1988), Visvanathan *et al.* (1997), Waring (1988), Pettman (1996) and Moon (1995) provide general background about women and development. Other works that development practitioners and impact assessment practitioners working in developing countries should consult include Chambers (1983, 1993, 1997) and Hancock (1989).

References

Banuri, T and Holmberg, J (1992) *Governance for Sustainable Development: A Southern Perspective*, International Institute for Environment and Development, London

Bass, S, Dalal-Clayton, B and Pretty, J (1995) *Participation in Strategies for Sustainable Development*, International Institute for Environment and Development, London

Branch, K and Ross, H (1997) The evolution of Social Impact Assessment: conceptual models and scope, Paper presented to the annual meeting of the International Association for Impact Assessment, New Orleans

Burdge, RJ (1994) *A Community Guide to Social Impact Assessment*, Social Ecology Press, Middleton

Burdge, RJ and Vanclay, F (1995) Social impact assessment, in *Environmental and Social Impact Assessment*, Vanclay, F and Bronstein, DA (eds), John Wiley, Chichester, pp. 31–65

Burkey, S (1993) *People First: A Guide to Self-reliant, Participatory Rural Development*, Zed Books, London

Cernea, MM (ed.) (1991) *Putting People First: Sociological Variables in Rural Development* 2nd edn, Oxford University Press, New York

Cernea, MM (1993) Sociological work within a development agency: experiences in the World Bank, unpublished paper

Cernea, MM and Guggenheim, SE (eds) (1993) *Anthropological Approaches to Involuntary Resettlement: Policy, Practice and Theory*, Westview, Boulder

Cernea, MM and Kudat, A (eds) (1997) *Social Assessments for Better Development: Case Studies in Russia and Central Asia*, The World Bank, Washington DC

Chambers, R (1983) *Rural Development: Putting the Last First*, Longman, Harlow

Chambers, R (1993) *Challenging the Professions: Frontiers for Rural Development*, Intermediate Technology Publications, London

Chambers, R (1997) *Whose Reality Counts: Putting the First Last*, Intermediate Technology Publications, London

Cohen, E (1983) Insiders and outsiders: the dynamics of development of bungalow tourism on the islands of southern Thailand, *Human Organisation* **42**: 158-162

Cohen, E (1984) The sociology of tourism, *Annual Review of Sociology* **10**: 373–392

Edensor, T (1998) *Tourists at the Taj: Performance and Meaning at a Symbolic Site*, Routledge, London

Engel, P and Salomon, M (1997) *Facilitating Innovation for Development: A RAAKS Resource Box*, KIT Press (Royal Tropical Institute), Amsterdam

Feldstein, HS and Jiggins, J (eds) (1994) *Tools for the Field: Methodologies Handbook for Gender Analysis in Agriculture*, Intermediate Technology Publications, London

Fuggle, R (1995) Integrated environmental management in South Africa, in *Involving People in the Management of Change Towards A Sustainable Future* Wood, C, Wynberg, R and Raimondo, J (eds), Proceedings of the 15th annual IAIA meeting June 26–30, Durban, South Africa, International Association for Impact Assessment, Fargo, pp. 61–68

Guijt, I and Shah, M (eds) (1998) *The Myth of Community: Gender Issues in Participatory Development*, Intermediate Technology Publications, London

Hancock, G (1989) *Lords of Poverty* Arrow, London

Holland, J and Blackburn, J (eds) (1998) *Whose Voice: Participatory Research and Policy Change*, Intermediate Technology Publications, London

Interorganizational Committee on Guidelines and Principles for Social Impact Assessment (1994) Guidelines and principles for social impact assessment, *Impact Assessment* **12**: 107–152

de Kadt, E (1979) *Tourism: Passport to Development?*, Oxford University Press, New York (in conjunction with the World Bank and UNESCO)

MacDonald, M (1994) What's the difference: a comparison of EA in Industrial and Developing Countries, in *Environmental Assessment and Development: A World Bank – IAIA Symposium*, Goodland, R and Edmundson, V (eds), World Bank, Washington DC, pp. 29–34

McLaren, DE (1994) Public Involvement in Environmental Assessment of Tourism, in *Environmental Assessment and Development: A World Bank – IAIA Symposium* Goodland, R and Edmundson, V (eds), World Bank, Washington DC, pp. 114–124

Mathieson, A and Wall, G (1982) *Tourism: Economic, Physical and Social Impacts*, Longman, London

Moon, G (1995) *Free Trade: What's in it for Women?*, Community Aid Abroad, Melbourne

Munt, M and Mowforth, M (1998) *Tourism and Sustainability: New Tourism in the Third World*, Routledge, London

Nabane, N (1995) Lacking Confidence? A gender-sensitive analysis of CAMPFIRE in Masoka Village Wildlife and Development Series Booklet No 3, International Institute of Environment and Development, London

Overseas Development Administration (1995) *A Guide to Social Analysis for Projects in Developing Countries*, HMSO, London

O'Rourke, D (1980) *Yap: How did you know we'd like TV* (video), Ronnin Films, Canberra

Peiris, K (1997) *Weaving a Future Together: Women and Participatory Development*, International Books, Utrecht

Pettman, JJ (1996) *Worlding Women: A Feminist International Politics*, Allen and Unwin, Sydney

Phillip, C (1983) *Assault on Paradise: Social Change in a Brazilian Village*, Random House, New York

Pretty, JN, Guijt, I, Thompson, J and Scoones, I (1995) *A Trainer's Guide for Participatory Learning and Action*, International Institute for Environment and Development, London

Roe, D, Leader-Williams, N and Dalal-Clayton, B (1997) *Take Only Photographs, Leave Only Footprints: The Environmental Impacts of Wildlife Tourism*, International Institute of Environment and Development, London

Royal Tropical Institute (1998) *Gender Training: The Source Book*, KIT Publications, Amsterdam

Salmen, LF (1987) *Listen to the People: Participant-Observer Evaluation of Development Projects*, Oxford University Press, New York

Shiva, V (1988) *Staying Alive: Women, Ecology and Development*, Zed Books, London

Slocum, R, Wichhart, L, Rocheleau, D and Thomas-Slayter, B (eds) (1995) *Power, Process and Participation*, Intermediate Technology Publications, London

Taylor, CN, Bryan, CH and Goodrich, CG (1995) *Social Assessment: Theory, Process and Techniques* 2nd edn, Taylor Baines and Associates, Christchurch

Truong, T (1990) *Sex, Money and Morality: Prostitution and Tourism in South-East Asia*, Zed Books, London

Vanclay, F (1999) Social Impact Assessment, in *International Handbook of Environmental Impact Assessment*, Petts, J (ed.), Blackwell Science, Oxford, pp. 301–306

Visvanathan, N, Duggan, L, Nisonoff, L and Wiegersma, N (eds) (1997) *The Women, Gender and Development Reader*, Zed Books, London

Wahab, S and Pigram, JJ (eds) (1998) *Tourism, Development and Growth: The Challenge of Sustainability*, Routledge, London

Waring, M (1988) *Counting for Nothing: What Men Value and What Women are Worth*, Allen and Unwin, Sydney

World Bank (1995a), *Social Assessment* Environment Department Dissemination Notes, Number 36, World Bank, Washington DC

World Bank (1995b), *World Bank Participation Sourcebook*, World Bank, Washington DC

8

Reviewing the Quality of Environmental Assessments

Norman Lee

8.1 Introduction

The most visible document produced during the EIA process is the environmental impact statement (EIS). This chapter is mainly concerned with assessing its quality. This should not be confused with the quality or acceptability of the project to which the EIS relates. The chapter also considers, but more briefly, how quality reviewing may be extended to cover:

a) The performance of the EIA process as a whole
b) The quality of SEA reports and of other new forms of appraisal reports which may be prepared in the future

The chapter is mainly concerned with the methodological aspects of quality reviews. However, procedural and institutional factors are also important and should be borne in mind throughout. For example, the quality and effectiveness of an EIS review also depends on:

- The stage in the EIA process at which it is undertaken
- The qualifications, experience and degree of independence of the reviewers
- The availability of the relevant documentation for review
- The resources and time provided for review
- The transparency and degree of participation in the reviewing process

Environmental Assessment in Developing and Transitional Countries. Edited by N. Lee and C. George.
© 2000 John Wiley & Sons, Ltd.

- The status of the review findings
- The use made of these at subsequent stages of the EIA process and project cycle

8.2   Assessing EIS Quality: Context, Criteria and Method

The quality of EISs has to be assessed taking into account the regulatory and procedural context in which they are prepared. For example:

- Some countries and funding agencies distinguish between full or detailed EISs and preliminary or simplified EISs, and their information requirements differ
- Most countries and funding agencies indicate, in their regulations or guidelines, the types of information which an EIS should cover. Most require similar information but, in certain cases, there are important differences between them (e.g. relating to the coverage of socio-economic impacts)
- Some require that the assessment findings are reported in a self-contained document whilst others only require that they form part of the overall planning documentation
- Countries and agencies differ in the types of project authorization procedures into which the EIA process are to be integrated and in the stage or stages in the project cycle at which the EIS is to be submitted. This may influence the scope of the EIS, the range of alternatives which may be realistically considered, and the precision with which impact predictions may be prepared

Despite these differences, there is a growing consensus about what constitutes a good EIS, based on good assessment practice. Reviewers have identified similar tasks to be undertaken at each stage of EIA preparation and a similar range of assessment methods to undertake these tasks, as well as similar criteria by which a selection should be made between them (see, for example, Canter and Sadler 1997).

Hence, a good EIS is one which presents, in a form appropriate to its intended users, findings covering all assessment tasks employing appropriate methods of information collection, analysis and reporting. In particular, these should be used to:

- Describe the project and the base-line environmental conditions which it may influence
- Predict the magnitude and significance of the expected changes to those environmental conditions
- Identify and assess the impacts of alternatives which have been investigated and of any mitigatory measures which are proposed for inclusion in the project
- Present the findings in an appropriate form for the intended users of the statement

Based on this common understanding of good practice, it is possible to construct a review checklist or package (which can be adjusted to different regulatory and procedural conditions) to assist in the systematic and objective review of EIS quality. This may be used by a variety of stakeholders in the EIA process:

- By the proponent of the project and his consultants, as a quality check on the EIS before it is formally submitted to the authorities
- By or on behalf of the competent authority or funding agency approving the project
- By an independent commission, environmental authorities or NGOs involved in the review and consultation process

8.3   Review Checklists and Packages

A number of non-mandatory checklists and packages have been prepared to assist in EIS quality reviews. These include the Lee-Colley package (Lee and Colley 1992; Lee *et al.* 1999), the Oxford Brookes review framework (Glasson *et al.* 1996), the Dutch EIA Commission checklist (Ministry of Housing, Spatial Planning and the Environment 1994a,b), the European Commission's review checklist (European Commission 1994) and USEPA's EIS Rating System (Canter 1996, pp. 25 *et seq.*). In each case these were initially developed with the requirements of specific countries in mind, although certain of them have subsequently been adapted for use in other countries as well.

The main features of the Lee-Colley package are described below before presenting the findings of a number of EIS quality studies in the next section of the chapter. According to this method, the quality review of an EIS involves evaluating how well a number of assessment tasks have been performed which are grouped hierarchically into sub-categories, categories and areas. The areas and categories are summarized in Box 8.1.

Box 8.1   Assessment Areas and Categories in the Review Package

Areas	*Categories*
1. Description of the development, local environment and base-line conditions	Purpose, physical characteristics, scale and design of project; its land requirements; types and quantities of residuals (e.g. wastes) and methods or routes of their disposal; likely geographic extent of the affected environment; expected base-line environmental conditions in the absence of the project
2. Identification and evaluation of key impacts	Identification of potential impacts of the project; scoping of impacts; prediction of impact magnitude; assessment of impact significance
3. Consideration of alternatives and mitigation measures	Feasible alternatives considered, their environmental implications and reasons for their rejection; scope and effectiveness of proposed mitigative measures; developer's commitment to implement these measures
4. Communication of assessment findings	Layout of EIS: accessibility of its contents to non-specialists; avoidance of bias; information included on data sources and methods of analysis used; inclusion of a sufficiently comprehensive non-technical summary

---

Each sub-category, category and area task is to be separately assessed in sequence, culminating in an overall assessment of the EIS as a whole, using the following rating system:

A   Generally well performed, no important tasks left incomplete.
B   Generally satisfactory and complete, only minor omissions and inadequacies.
C   Can be considered just satisfactory, despite omissions and/or inadequacies.
D   Parts are well attempted but must, as a whole, be considered just unsatisfactory because of omissions and/or inadequacies.
E   Not satisfactory, significant omissions or inadequacies.
F   Very unsatisfactory, important tasks poorly done or not attempted.
N/A Not applicable. The Review Topic is not applicable as it is irrelevant in the context of this EIS.

The review findings are recorded on a one page Summary Sheet to assist in their analysis and interpretation.

A desk review of a relatively straight-forward EIS of 50–100 pages can be completed in 3–6 hours although a site visit, which is often desirable, would add to this time. In the case of more complex projects, this review procedure can also be used to identify specific topics where the advice of one or more specialist technical reviewers may be desirable. In all cases, where possible, each EIS should be separately reviewed by two people and their findings compared to reduce subjectivity in the reviewing process.

8.4   Findings of EIS Quality Reviews

A growing number of studies of EIS quality have been completed, particularly relating to high income countries but also in some low and middle income countries. Some of these, which used variants of the Lee–Colley Package, are reviewed below.

High Income Countries

In 1996, the European Commission published the findings of a study relating to EIS quality in eight Member States of the European Union – Belgium, Denmark, Germany, Greece, Ireland, Portugal, Spain and the UK (European Commission 1996). These showed that, on average, approximately 70% of the EISs reviewed were of satisfactory quality (i.e. 'C' grade or above). This is a significantly higher percentage than for a corresponding sample of EISs produced at the beginning of the 1990s (when it was approximately 50%) and higher still than in the late 1980s when mandatory EISs were first required in these countries.

An analysis of the Summary Sheets for the EISs reviewed in this study enables the principal shortcomings in their quality to be located more precisely. The assessment areas where performance is least satisfactory in high income countries are: Area 2 (Identification and Evaluation of Key Impacts) and Area 3 (Alternatives and Mitigation). In the former case, this is mainly due to technical and other weaknesses in scoping, impact prediction and determination of impact significance. In the latter case, it is often due to insufficient consideration of alternatives and insufficient commitment by the developer or promoter to the mitigation measures which are proposed. Area 1 tasks, relating to the description of the project and the base-line environment, are relatively better performed but the quantities of wastes associated with the projects are less satisfactorily estimated. Similarly, Area 4 tasks are performed better than average, especially in the more recent EISs, but the quality is lower in a number of cases due to biased reporting and inadequate non-technical summaries.

Developing Countries

In 1992, Ibrahim analysed the quality of 13 EIA reports submitted to the Department of Environment in Malaysia between 1988 and 1991 (Ibrahim 1992). These were selected to reflect the broad range of projects submitted to EIA in Malaysia at that time. Of these, 8% were assessed as good quality (A or B ratings), 77% were borderline (C or D ratings) and 15% were poor (E or F ratings). Overall, the quality of the reports was best in Review Area 4, followed by Area 1 and then Areas 2 and 3, respectively.

In 1993, Rout reviewed the quality of 7 EIA reports submitted to the state authorities in Orissa, India (Rout 1994). He found approximately 30% to be satisfactory (all in the C category) and 70% were unsatisfactory (in D, E or F categories). The quality of the reports was best in Area 1, followed by Area 4, and was least satisfactory in Area 3 and Area 2.

In 1997, Mwalyosi and Hughes used a simplified version of the review package to assess the quality of 26 EISs prepared in Tanzania (Mwalyosi and Hughes 1998). Their ratings in the main review areas and categories are higher than in the two earlier studies but their accompanying commentary is more critical. The quality of the reports was considered best in Area 1 (73% had A or B ratings) followed by Area 2 (42% with A or B ratings) and Area 3 (31% with A or B ratings). Their general observation is that 'the EISs tend to be descriptively strong, but analytically weak' (p. 37). The rating percentage is not provided for Area 4 but, in the commentary, the presentation of findings is assessed as often poor. 'In only one third of cases were recommendations in EISs sufficiently clear that proponents could reasonably be expected to use them. Less than half of all EISs contained clear and comprehensive executive summaries that provide suitable information for decision-making.' (p. 39).

The more detailed results of these three studies can be used to identify the main categories within the four review areas where the greatest deficiencies in EIS quality occur. As can be seen from Box 8.2, there are many similarities between their findings.

Comparisons between reviews in EIS quality in widely differing countries, developed and developing, have to be made with considerable caution. However, the findings are sufficiently similar for the following broad conclusions to be drawn:

- In both developing and developed countries, there is clear evidence of unsatisfactory quality EISs being produced and for this to be a source of concern
- The review areas and categories where quality problems are the greatest have been identified and are seen to be very similar in most countries, both developing and developed. This is helpful to know when formulating priorities for improving EIS quality
- There is also evidence of good quality assessments being undertaken in particular EIS review categories and of overall quality improving as experience grows in carrying out EISs especially where it is supported by institution strengthening and appropriate guidance and training. This experience is helpful when developing specific strategies and measures to improve EIS quality

---

Box 8.2   Assessment Categories where EIS Quality is Least Satisfactory

Ibrahim (1992)	Estimation of residuals; scoping; assessment of magnitude and significance of impacts; consideration of alternatives; commitment to mitigation measures
Rout (1994)	Estimation of residuals; identification and scoping of impacts; assessment of significance of impacts; consideration of alternatives; commitment to mitigation measures; bias; non-technical summary
Mwalyosi and Hughes (1998)	Estimation of residuals; assessment of impact significance; consideration of alternatives; scope and effectiveness of mitigation measures; presentation of results, including gaps and uncertainties in information provided

---

8.5   Assessing the Performance of the EIA Process

In the final analysis, it is the performance of the EIA process as a whole, rather than the quality of the EIS, which is of importance. A good quality EIS cannot guarantee a high performing EIA process, although it should contribute to this.

Assessing the performance of the process is more complex and less well-developed than assessing EIS quality. Two broad approaches are being followed (Lee *et al.* 1994):

a) Aggregate analyses which assess the benefits and costs resulting from the process as a whole. These attempt to determine: (i) the number of projects which are modified and the scale of environmental improvements which are attributable to the EIA process; and (ii) the financial, time and other costs which are incurred due to the process
b) Disaggregate analyses which use indicators of process effectiveness or outcomes (positive and negative) at individual stages in the EIA process. These indicators relate to: the timing of the beginning of the EIA process relative to the commencement of the project cycle, the degree of compliance with good practice scoping requirements, the quality of the EISs which are submitted, the extent

and effectiveness of public participation, the degree of integration of EIA find-ings into authorization decisions and the extent to which the environmental conditions of those decisions are implemented

The broad conclusions of the aggregate studies relating to developed countries show mixed results (see, Lee *et al*. 1994; European Commission 1996; CEC 1993; and Sadler 1996). Where EIA regulations are well-formulated and function effi-ciently, significant environmental benefits are being realized at little additional cost. For example, in the UK and certain other EU countries:

- Half or more of development projects subject to EIA are modified to reduce their negative environmental impacts
- The costs of EIS preparation are typically around 0.2% of total project cost and only exceptionally increase above 1% of this capital cost
- In well-managed systems, particularly where a good quality EIS has been pro-duced, the average time to obtain a project authorization may decline

However, the same studies also show:

- A high proportion of project modifications may be made at a late stage in the project cycle and are then of relatively minor environmental significance
- Where the developer starts his EIA process late in the project cycle, and where the consultation, public participation and decision-making activities of the au-thorities are not well organized, the incremental costs and delays for all parties can be considerable

The balance between these favourable and unfavourable circumstances in de-veloped economies varies between national EIA systems and over time. Some more detailed information is available from disaggregate studies relating to individ-ual stages of the EIA process (e.g. Kobus and Lee 1993).

Less information is available relating to the performance of the EIA process in developing countries and countries in transition (Biswas and Agarwala 1992; Sadler and Verheem 1997; Bellinger *et al*. 1999). Again the picture is a mixed one, but usually with greater emphasis on shortcomings in both regulatory provisions and actual practice.

Ibrahim's 1993 study of EIA performance in Malaysia identifies a number of positive features but also indicates:

- EIA reports were often prepared and submitted too late in the project cycle. In the period 1988–1990, one-third of EIA reports were submitted after the feasi-bility stage had been completed. Also, 45% of projects commenced ground preparation before their EIA reports had been approved
- A significant proportion of consultants engaged in EIA work had no previous EIA experience and criteria for their evaluation were not in place
- Due to a sharp increase in the number of EIA reports, without a corresponding increase in Ministry staff, significant delays in reviewing reports occurred

- At that time, impact and compliance monitoring of projects subject to EIA was only in its infancy and had not been developed for most projects

Rout's 1994 review of the performance of the EIA process in Orissa, distinguished between pre- and post-EIA report submission phases:

- The pre-submission phase tended to start too late in the project cycle. Frequently EIA reports were prepared during or after the project was being implemented. The degree of integration between the EIA process and project planning and design was often limited. Although the experience of EIA consultants was growing, the quality of their studies was reduced by the late stage in project planning and design at which they were engaged to commence work
- At the time there was no formalized provision for public participation in either phase of the process. Reviewing EIA reports was an integral part of the process, as in Malaysia, but was not a source of delay. In some cases, the review findings were influencing siting decisions and mitigation requirements. However, the EIA process requirements contained no arrangements for monitoring compliance but this may be partly covered under separate regulatory provisions

The EIA performance study in Tanzania by Mwalyosi and Hughes (1998), based on seven case studies, acknowledged certain benefits but was critical of the overall performance of the EIA process:

> Our findings show that EIA has had very little impact on decision-making in Tanzania. In most cases, EIAs were extremely late in starting, under-resourced and generally omitted to involve other stake-holders to any meaningful extent . . . .. In most cases, the EIS did not define, cost and integrate environmental management into project design, and few defined compliance responsibilities. Perhaps, not surprisingly, compliance with recommendations of EIA has been the exception rather than the rule. Consideration of alternative project options was often absent, or extremely weak . . . .. The study found no evidence that donor agency-supported processes . . . led to more effective EIA . . . .. The study found no examples where donor agency interest extended to ensuring EIA recommendations were adhered to during implementation, post completion or audit phases of the project (pp. 73–74)

The types of deficiencies which have been identified in the performance of the EIA process in developing countries are attributed to a wide range of factors. These include: deficient regulatory provisions, limited experience in their application due to their recent enactment; the complexities in the EIA process associated with the conflicting assessment requirements of overseas funding agencies and national authorities; limited in-country capacities to organize, manage and carry out environmental assessments and to integrate them successfully within the project cycle; political and cultural obstacles to implementing certain components of EIA practice, notably those requiring public participation; and general weaknesses in enforcing environmental regulations in circumstances where economic development is given the highest priority (see Chapter 10 for further details).

These deficiencies need to be more fully researched in order to construct more appropriate measures to improve EIA performance in developing countries.

However, it is already clear that the range of measures required extends well beyond those needed to improve EIS quality.

8.6   New Forms of Assessment and Quality Assurance

As new forms of assessment emerge, quality review systems will need to adapt to these. Three kinds of development in environmental assessment practice can already be envisaged:

- The extension of environmental assessments to policies, plans and programmes (i.e. strategic environmental assessment)
- A growing interest in sustainability issues within environmental appraisals
- The increasing use of integrated appraisals (combining economic, social and environmental assessments) in promoting sustainable development

Quality Reviews of SEA Reports

Work has already commenced on the development of review packages to check the quality of SEA reports relating to policies, plans and programmes. For example, the Lee–Colley package has been adapted by others to check the quality of environmental appraisals of land use plans (Lee *et al.* 1999). This new package is similar in structure to the project-level package but differs in its detailed content in order to reflect the more strategic character of appraisals required at earlier stages in the planning cycle.

Some preliminary tests have been completed using the new package to evaluate the quality of small numbers of environmental appraisals of land use plans in the UK and Sweden (Lee *et al.* 1999). The initial findings suggest that sizeable proportions of these appraisals are not of a satisfactory standard. This corresponds to the experience with project-level EISs where significant deficiencies were found in the quality of the majority of first generation statements. Also, interestingly, the most deficient review areas appear to be the same – Area 2 (Identification and Evaluation of Key Impacts) and Area 3 (Consideration of Alternatives and Mitigation Measures) although deficiencies are to be found in all four review areas.

The main deficiencies which have been highlighted, so far, are:

- The environmental appraisal documentation is not sufficiently self-contained and user friendly
- Base-line environmental conditions, in the absence of plan implementation, are not adequately determined or sufficiently distinguished from existing environmental conditions
- The determination of the scope of the environmental appraisal, the prediction of the plan's impacts and the determination of their significance are not undertaken in a sufficiently systematic and explicit manner
- The environmental appraisal of the main alternatives to the plan's proposals is not adequately presented nor is the selection of the recommended proposals sufficiently justified in environmental terms

It is important that the lessons to be drawn from assessing the quality of SEA reports are brought to the attention of both authorities and practitioners in developing countries and countries in transition as soon as possible. In this way, they can be assisted in developing their own quality assurance initiatives in the early stages of introducing SEA processes in their own countries.

Quality Reviews and Sustainability

The increased interest in sustainability issues is beginning to be reflected in the content of both SEA and EIA reports. In many cases, this is not yet being done satisfactorily, often due to the absence of well-defined and properly justified sustainability indicators (Dalal-Clayton 1992; Moldan and Billharz 1997; Shillington *et al.* 1997). Paralleling this, there is a need to extend the criteria used in review checklists and packages to cover more explicitly the treatment of:

a) The consumption of natural resources and the damage caused to them due to proposed actions
b) The compatibility of these impacts with appropriately defined sustainability targets

At present, the development of such review criteria is still in its infancy.

Quality Reviews and Integrated Appraisals

Another interesting initiative is the development of integrated appraisals of PPPs and projects which attempt to bring together their likely economic, social and environmental consequences within a single evaluative framework (Kirkpatrick and Lee 1997; Lee and Kirkpatrick 1999). These are essentially 'sustainable development' appraisals whose importance is likely to grow in the future in both developed and developing countries. The means by which these integrated appraisals will be realized, or the form they will finally take, is not yet fully clear. One approach, for example, is to integrate economic, social and environmental appraisals throughout the planning and project cycle. In this case, the review package or checklist will also need to adopt an integrated form. Alternatively, each type of appraisal may retain its separate identity and only be brought together at key decision points in the planning and project cycle (e.g. in an extended cost-benefit analysis or multi-criteria decision framework). In this case, the review package would take a different form and its scope could vary according to the stages in the planning and project cycle at which it is to be used.

8.7   Conclusions

In summary, the means have been developed, and are now being more widely used, for reviewing the quality of EISs in both developed and developing countries. Their use (by developers, authorities, funding agencies, NGOs and other consultees), in the detection of weaknesses in key environmental assessment documentation, is important to the improvement of future assessment practice at the project level.

However, the practice of quality reviews needs to be extended beyond these boundaries. First, it needs to be broadened to cover the overall performance of the EIA process and its constituent stages. Secondly, in the near future, reviews will need to cover SEA reports and the overall performance of the SEA process. Finally, reviews will eventually need to cover sustainability issues both within environmental appraisals and, possibly, in broader integrated appraisals.

Given the limited capacities and resources in many developing countries for impact appraisals and their quality control, these extensions to reviewing practice will need to be introduced in a sufficiently simplified and practical manner, probably using a step-by-step approach in the first instance.

Discussion Questions

1. What are the main qualities of a good Environmental Impact Statement? How might the overall quality of an EIS be best assessed?
2. What are the main deficiencies in (a) the quality of EISs and (b) the overall performance of the EIA process, in low and middle income countries? What are the main measures you would propose to improve this situation?
3. What are the main directions in which quality reviews of environmental assessments should develop in the future? What are the main difficulties to be overcome in extending quality reviews in these directions?

Further Reading

Additional information on quality review checklists can be found in Lee *et al.* (1999), European Commission (1994), Ministry of Housing, Spatial Planning and the Environment (1994a,b) and Canter (1996). Findings of EIS quality reviews in different types of countries are contained in Mwalyosi and Hughes (1998), Biswas and Agarwala (1992), Sadler and Verheem (1997) and Lee *et al.* (1999). Quality reviews of SEA reports are discussed in Lee *et al.* (1999). Information relevant to the development of review criteria to appraise sustainability appraisals and integrated appraisals are to be found in Dalal-Clayton (1997), Kirkpatrick and Lee (1997) and Lee and Kirkpatrick (1999).

References

Bellinger, E, Lee, N, George, C and Paduret, A (eds) (1999) *Environmental Assessment in Countries in Transition*, Central European University Press, Budapest

Biswas, AK and Agarwala, SBC (1992) *Environmental Impact Assessment for Developing Countries*, Butterworth-Heinemann, Oxford

Canter, LW (1996) *Environmental Impact Assessment*, 2nd edn, McGraw-Hill, New York

Canter, L and Sadler, B (1997) *A Tool Kit for Effective EIA – Review of Methods and Perspectives on their Application*, Environmental and Groundwater Institute, University of Oklahoma, Norman, OK

Commission of the European Communities (CEC) (1993) *Report from the Commission of the Implementation of Directive 85/337/EEC* COM 93(28), Commission of the European Communities, Brussels

Dalal-Clayton, B (1992) *Modified EIA and Indicators of Sustainability: First Steps Towards Sustainability Analysis*, International Institute of Environment and Development, London

European Commission (1994) *Environmental Impact Assesment Review Checklist*, European Commission (DGXI), Brussels

European Commission (1996) *Evaluation of the Performance of the EIA Process* 2 vols, European Commission (DGXI), Brussels

Glasson, J *et al.* (1996) *Changes in the Quality of Environmental Statements for Planning Projects*, Department of the Environment, HMSO, London

Ibrahim, AKC (1992) An Analysis of Quality Control in the Malaysian Environmental Impact Assessment Process, unpublished MSc dissertation, University of Manchester, Manchester

Kirkpatrick, C and Lee, N (eds) (1997) *Sustainable Development in a Developing World: Integrating Socio-Economic Appraisal and Environmental Assessment*, Edward Elgar, Cheltenham

Kobus, D and Lee, N (1993) The role of environmental assessment in the planning and authorisation of extractive industry projects, *Project Appraisal* **8**: 147–156

Lee, N and Colley, R (1992) *Review of the Quality of Environmental Statements*, Occasional Paper Number 24, 2nd edn, EIA Centre, University of Manchester, Manchester

Lee, N and Kirkpatrick, C (eds) (1999) *Sustainable Development and Integrated Appraisal in a Developing World*, Edward Elgar, Cheltenham (in press)

Lee, N, Colley, R, Bonde, J and Simpson, J (1999) *Reviewing the Quality of Environmental Statements and Environmental Appraisals*, Occasional Paper 55, EIA Centre, University of Manchester, Manchester

Lee, N, Walsh, F and Reeder, G (1994) Assessing the performance of the EIA Process, *Project Appraisal* **9**: 161–172

Ministry of Housing, Spatial Planning and the Environment (1994a) *The Quality of Environmental Impact Statements*, VROM Series, No 47, VROM, Zoetemeer, The Netherlands

Ministry of Housing, Spatial Planning and the Environment (1994b) *Use and Effectiveness of Environmental Impact Assessment in Decision-Making*, VROM Series, No 49, Zoetemeer, The Netherlands

Mwalyosi, R and Hughes, R (1998) *The Performance of EIA in Tanzania: an Assessment*, International Institute for Environment and Development, London

Rout, DK (1994) An Analysis of the EIA Process and EIA Reports produced for selected industrial developments in the State of Orissa in India, unpublished MSc dissertation, University of Manchester, Manchester

Sadler, B (1996) *Environmental Assessment in a Changing World: Evaluating Practice to Improve Performance*, Final Report of the International Study of the Effectiveness of Environmental Assessment, Ministry of Supply and Services, Ottawa

Sadler, B and Verheem, R (1997) *Country Status Reports on Environmental Impact Assessment*, Netherlands Commission for Environmental Impact Assessment, Utrecht

Shillington, T, Russell, D and Sadler, B (1997) *Addressing Climate Change through Environmental Assessment: a Preliminary Guide*, Royal Society of Canada and Canadian Environmental Assessment Agency, Ottawa

9

Methods of Consultation and Public Participation

Ron Bisset

9.1   Why is Consultation and Public Participation Important?

Since the formal beginning of Environmental Assessment (EA) in the early 1970s, consultation and public participation (CPP) has been a feature of many national EA systems (Roberts 1995). In recent years, however, there has been an undoubted increase in the attention paid to this activity with national laws and regulations containing specific and detailed procedures for consultation and public participation (UNEP 1996). An examination of the EA procedures of the multi-lateral financing agencies such as the World Bank group, and bi-lateral aid agencies such as the Canadian International Development Agency and the UK Department for International Development, shows a parallel interest in ensuring that the public is involved in EA activities (Davis and Soeftestad 1995; Mutemba 1996; World Bank 1993). The question arises – why has this national and 'international' upsurge in concern, for enhancing the public input to EA, occurred?

The reasons can be divided into three categories: first, lessons learned from *ex post* project evaluations; secondly, reasons arising from political and policy changes at global and national levels; and, finally, those resulting from the evolving policies of the multi- and bi-lateral agencies:

Lessons from *ex post* Evaluations

- Better designed projects which avoid costly delays in appraisal and implementation result from early and planned CPP

Environmental Assessment in Developing and Transitional Countries. Edited by N. Lee and C. George.
© 2000 John Wiley & Sons, Ltd.

- Projects with CPP are often more likely to achieve their objectives
- Projects are less likely to fail if the public is involved (an example is provided in Pretty (1993))

Political and Policy Changes

- Increase in the number of countries with 'western-style', representative democratic forms of government (with a tradition of CPP, to varying degrees, in policy and project-related decision-making)
- Trend, in certain countries, toward decentralization of decision-making
- Trend towards a reduction in the influence of the public sector, accompanied by an increase in privatization
- Growth in influence of non-governmental organizations (NGOs) (UNEP 1996)

Policies of Multi- and Bi-laterals

- Promotion of good governance
- Emphasis on poverty alleviation and gender
- Promotion of capacity development (OECD 1994)

Probably, the trends at the global political level and in the policies of the multi- and bi-laterals have been more influential so far, than the lessons learned from evaluations. The overall implication of all these factors has been to move CPP closer to a central role in the development process and, hence, in EAs. Of course, as will be shown below, practice varies considerably and the successful and consistent integration of CPP within EA systems is not likely to be realized for many years yet.

9.2   What is Meant by Consultation and Public Participation?

In recent years, a consensus has begun to emerge on the definitions of the various terms used when CPP is discussed. Unfortunately, this 'agreement' has not yet been universally adopted. There are four main types of CPP which can be identified, based upon a review of this emerging consensus (see Box 9.1).

The term 'participation' refers, usually, to the latter two categories because in these cases the public has a direct and acknowledged role in decision-making. In global EA practice, consultation is probably the most common type of CPP, but there is an increasing interest in exploring the possibilities of the more participatory approaches. Box 9.2 shows two contrasting types of CPP in national EA procedures.

There is a definite momentum towards giving the public more control over the conduct of EAs. For example, the World Bank favours the 'collaboration and partnership' approach when the interests of indigenous peoples are likely to be affected (World Bank 1993). Indigenous peoples are identified, using special criteria devised by the Bank, and include groups such as hunter-gatherers and marginal and vulnerable communities such as scheduled tribes and castes in India. This approach is also used in cases where a project will cause involuntary resettlement.

Box 9.1   Types of CPP

Consultation applies to the involvement of national and local government agencies, whose interests may be affected directly by a proposed project, and of the public and NGOs. Emphasis is placed below on types of CPP involving the public and NGOs.

1. *Information Dissemination*

In this case, information is provided by the proponent, or an agency which is responsible for the EA, to the public on one occasion or at regular intervals. However, the flow of information is 'one way'. Usually, there is no opportunity for comment by the public or NGOs on the merits of the proposed project.

2. *Consultation*

This involves a 'two way' process of exchange of information between the public and a proponent/agency with the opportunity for the public to comment on the merits of the proposed project. Consultations may occur at varying times during the EA process. However, the proponent or agency is not required to take account of the views expressed in its decision-making, although it may do so if it considers it appropriate.

3. *Collaboration and Partnerships*

In this case, the public is considered to be a partner in the development initiative and in carrying out the EA. There is shared decision-making and control in the EA process and joint responsibility for the EA results. There is joint analysis and control over decisions and their implementation. The partnership may extend to the design and implementation of the proposed project.

4. *Empowerment and Local Control*

In this case, control over the scope, form and content of the EA is passed to the local community(ies) and it is exercised, usually, through community representatives.

An additional, important trend in CPP can be identified. It has been understood, for some time, that the 'public' is not a homogenous body with a set of agreed common interests and aims. Rather it is a mix of different interests which are often conflicting. This realization has led to increasing use of the concept of stakeholders. Stakeholders are social groups and categories, organizations, agencies and individuals whose interests may be beneficially or adversely affected by a proposed project. Increasingly, the term 'public' is being replaced by that of 'stakeholders' in EA practice and this usage will be followed below.

9.3   How to Implement CPP?

The first step should be to undertake a stakeholder analysis to identify those to be involved in an EA. Some stakeholders are easy to identify – for example, the proponent, the government ministry or agency in charge of the EA procedures and central and local government organizations whose remit and responsibilities include areas and sectors likely to be affected (such as health, natural resources and land use). It is essential that these stakeholders are included in EA consultations. Unfortunately, it is still the case that intra-government consultations are weak in

---

Box 9.2   CPP in EA Procedures in Moldova and Zimbabwe

Moldova

In Moldova the public can undertake its own assessment of the documentation (including the EA Report if one has been prepared) relating to a proposed project under the Ecological Expertise system, but only if the individuals concerned form an association. This must meet certain criteria (for certain types of projects the association has to include at least 100 people) before the obligatory registration with a local government entity can be realized. NGOs may constitute such an association. Individual citizens may send written comments and/or objections to the authorities which should be taken into account in decision-making, but at the discretion of the competent authorities. National and local government organizations, also, are consulted. In Moldova, CPP is restricted to commenting on documents, usually produced as a result of an assessment, and apart from provision of written comments from individuals, CPP is restricted to approved associations and government bodies as indicated above.

Zimbabwe

CPP requirements vary with the scale of a project and its location. Minimum requirements are specified but the Zimbabwe EA Policy indicates that '. . . more problematic activities should involve more extensive consultation'. CPP should encompass all stakeholders and, according to the accompanying EA Guidelines, be initiated as early as possible in the project cycle and occur at the following main stages:

- Scoping and preparation of the EA Terms of Reference
- Preparation of the EA Report
- Government review of the EA Report
- The preparation of terms and conditions for EA Acceptance (conditions attached to a permission enabling a proposed project to be implemented)

The results of the CPP are to be documented in the EA Report, but the proponent and competent government authorities are not under any obligation to take account of them in decision-making.

---

many countries despite legal requirements for consultation. Similarly, it can be relatively easy to identify certain NGOs and community-based organizations (CBOs), working in the local area, which should be involved. Problems can arise, however, in identifying international environmental NGOs which may not be working locally, but may be based in the capital.

It is less easy to identify the other local stakeholders. Consultation between the proponent, government and those undertaking the EA may be necessary to establish, preliminarily, the spatial boundaries of the EA and to identify local individuals, social groups/categories and organizations who are local stakeholders. Stakeholders will vary according to the EA and the locality involved. Experience indicates that the main types of local stakeholders, not including local offices of central government ministries/agencies and local governments and NGOs/CBOs, are:

- Project beneficiaries (who *may or may not* be local)
- Local communities (which may be single villages or groups of villages which share a common politico-legal allegiance or framework and/or cultural identity).

In multi-ethnic situations a number of culturally distinct communities may need to be defined
- Voluntary organizations such as local community development or resource user groups, kinship societies, recreational groups, neighbourhood associations, gender based groups, labour unions and co-operatives
- Private sector bodies such as professional societies, trade associations and chambers of commerce
- Indigenous peoples (if present)
- Non-resident social groups who may use local resources either regularly or intermittently, for example pastoralists (likely to be classed as indigenous people) or tourists
- Selected social categories, for example, women, the elderly (especially if resettlement is an issue) and the poorest people

This is not an exhaustive list; it merely indicates the variety and types of local stakeholders which might be involved most often in EAs. In certain contexts it may also be considered useful to involve individuals, such as academics and research scientists, with special local knowledge or skills.

Once the stakeholders have been identified it is possible to begin the process of involving them in the EA. Current EA practice first involves stakeholders in scoping activities to:

- Determine the time and space boundaries for the EA (this may be derived from a preliminary definition used to identify the stakeholders)
- Identify whether any additional stakeholders should be involved
- Identify feasible alternative project designs and locations
- Identify the impacts to be investigated in the EA
- Agree a plan for future stakeholder involvement (this is not always a component of scoping)
- Provide information, in the form of indigenous knowledge, for those carrying out the EA
- Agree the outline of the Terms of Reference (ToR) for the EA

Once the ToR has been completed, EA investigations begin. Generally, stakeholder involvement does not occur again until a draft or final EA Report is available (unless it is the type of CPP which involves collaboration or control). The availability of these Reports, the language in which they are written, the time period for comment and the extent of accessibility are all important issues which influence the effectiveness of CPP. In most countries, with CPP provisions, comments on these Reports are solicited and then may be taken into account in the decision-making process.

However, depending on circumstances and opportunities, or if a plan for stakeholder involvement has resulted from scoping, it is possible to extend stakeholder involvement to:

- Various other stages in the EA, for example when preliminary or interim EA Reports become available

- Project implementation (application of EA recommendations) following project approval
- Project evaluation (determining the extent to which the project has achieved its objectives and undertaking a performance review of the EA which has been carried out)

Effective stakeholder involvement in the EA process requires that the following difficulties and constraints, which characterize many low and middle income countries, have to be addressed:

- Illiteracy
- Linguistic and cultural diversity which hinder mutually intelligible communication
- Lack of local knowledge and understanding regarding the scale, nature and likely effects of certain types of development projects
- Unequal access to consultative and participatory processes for certain social categories (for example, women)
- Remoteness of some stakeholders
- Time/cost implications of dealing satisfactorily with these difficulties

It has been tempting, in the preparation of some EAs, to rely entirely on NGOs/CBOs to act as representatives for local communities, but there are considerable dangers in this 'solution' despite the resource savings that may result. First, they are *not* able to reflect the extent of the diversity in local interests and, secondly, the time they spend acting as community representatives can detract from their own CPP activities. Finally, the local community groups should be treated as stakeholders, in their own right, because of their local knowledge and interests.

It is important that a plan for stakeholder involvement is prepared before EA work begins and this should preferably be part of the ToR. If this is not possible then a separate plan could be prepared subsequently, probably by the EA team leader, with input from an anthropologist or rural sociologist with knowledge of the local cultures and different approaches or techniques available for implementing stakeholder involvement. Such a plan should consider:

- The objectives of the EA
- The stakeholders to be involved
- Matching of stakeholders with approaches and techniques of involvement
- Traditional authority structures and political decision-making processes
- Programming of the implementation, in time and space, of the different approaches and techniques for stakeholder involvement
- Mechanisms to collect, synthesize, analyse and, most importantly, present the results to the EA team and key decision-makers
- Measures to ensure timely and adequate 'feedback' to the stakeholders
- Budgetary/time opportunities and constraints

A brief description is presented in Box 9.3 of a stakeholder involvement programme for a strategic EA of alternative development scenarios likely to affect an area within a 30 km radius of the Victoria Falls World Heritage Site on the border between Zambia and Zimbabwe (UNEP 1996). It illustrates the range of stakeholders and the different techniques used to obtain their input to the EA.

Box 9.3   Stakeholder Programme: Victoria Falls Strategic EA

The programme was aimed, primarily, at local residents (in their various communities and groups), NGOs with a local base, the private sector, tourists and national and local government agencies. The programme was devised by the EA team leaders assisted by two anthropologists who spoke the main local languages. Throughout the EA there was a media campaign involving the local press, radio and TV to raise public awareness and to keep them informed of the progress of the EA.

Two broad categories of residents were defined as needing different techniques. First, were the communities, mostly village-based under traditional authority with well-established institutions and customary decision-making procedures. These were used as a mechanism to involve residents. It was not possible to involve all villagers; therefore a representative sample of villages was identified and involvement was undertaken via the chiefs or headmen. In a number of cases, open public meetings were arranged to allow individuals to participate. Matrilineality occurred in one of the local cultures giving women a status enabling them to participate in public meetings. In addition, the anthropologists undertook random visits (with local political permissions) to villages and interviewed individual residents informally.

Secondly, there were urban residents including those living in Livingstone, Victoria Falls and the township housing the workers employed at the hydro-power station. These residents had many diverse interests and their social organization was correspondingly more complex. It was necessary to divide these residents into different social categories (for example, the poor living in informal settlements on the outskirts of Livingstone) and groups (for example, the association of curio sellers, various co-operatives and church and gender-based groups). The inhabitants of the informal settlements were visited by an anthropologist who undertook random, informal interviews. For each selected group, two or three representatives were identified and interviewed following a semi-structured questionnaire. Two open public meetings and a series of open houses were held in Livingstone and Victoria Falls to provide a further opportunity for local opinions to be obtained.

The private sector was involved by identifying trade associations (for example, local members of the Zimbabwe Council for Tourism) and the Chambers of Commerce. Representatives were interviewed in a semi-structured way. The views of tourists, visiting the Falls, were obtained by trained interviewers administering a questionnaire to a random sample. Finally, representatives of local and national governments were included as members of a Steering Group which met regularly to review work done and to guide future activities.

Feedback was achieved through public meetings (two meetings were held in Livingstone and Victoria Falls separated by a few months) and through the media.

There are numerous approaches and techniques which can be used to involve stakeholders. Amongst the most common are:

● *Public meetings.* These may be specially convened meetings of traditional local decision-making fora such as the *khotla* in Botswana or the *pitso* system in Lesotho (Kakonge 1996) which operate at village level. Their conduct is determined by customary rules. Public meetings may also be held in larger, urbanized communities and be conducted according to different sets of rules. Generally, they are 'open' with no restriction as to who may attend. Not everyone is comfortable speaking in such 'non-traditional' events so the views expressed may not be representative of the urban community

- *Advisory panels.* These consist of selected groups of individuals, chosen to represent stakeholders, which meet periodically to assess results and advise on future activities. The selection of the stakeholders and their representatives is crucial and particular care is needed to involve the poor, women and indigenous people if they are not already identified as stakeholders
- *Questionnaires.* A structured set of written questions is prepared to be answered by a sample of local people. A sufficient level of literacy in the local community is assumed since no interviewing may be undertaken. If interviewing is needed then the costs will rise
- *Interviews.* These may be structured or semi-structured. Usually, interviews are undertaken with selected community or group representatives. They can also be undertaken with individuals considered able to reflect the views of distinct social categories, for example, the poor who may not be organized in a defined social group. This technique requires skilled personnel and adequate funding
- *Open houses.* These are accessible locations which contain information on the proposed project and the EA. An EA team member should be present who can discuss issues and record the views/opinions/concerns of visitors. Any person should be able to visit an open house and the location and times of opening should be well publicized.

These different approaches should be used in varying combinations depending on local circumstances. They are used most often for consultation. For involvement relating to collaboration or control, there is less choice. Techniques may be selected from the 'toolkit' known as Participatory Appraisal Methods (Pretty *et al.* 1993), supported by one or more of the above consultation techniques to provide additional information and insights. Basically, the participatory techniques provide a systematic approach to appraisal founded on *group* inquiry and analysis. They provide varied and multiple inputs to the appraisal, but reaching a consensus based on these inputs is then very important. Assistance may be provided by external experts, but they are not expected to direct or control the appraisal.

9.4   Some Issues Relating to CPP

The implications of CPP vary according to whether consultation, collaboration or control is the aim. This section focuses on consultation; the implications of collaboration and control are discussed in the next section.

Adequate consultation takes time and resources and costs money. This is especially true for EAs implemented in remote locations with cultural diversity. Sufficient provisions have to be made in the budget to meet these costs. Also, sufficient time has to be allocated, within the EA timeframe, to enable the consultation to be planned and undertaken, the results to be analysed and used in the EA and feedback to be provided to the stakeholders. In particular, sufficient provision should be made for:

- EA team time, travel and accommodation costs to support the work (*especially the data collection, analysis and reporting work which is time-consuming and often neglected*)
- Hiring of social scientists with local knowledge and experience
- Travel costs and 'sitting allowances' to enable certain individuals to attend meetings
- Preparation of materials in local languages
- Management of media publicity

Since stakeholder involvement requires additional expenditure it is important to find the most cost-effective and efficient means of integrating stakeholder involvement in EAs. This is particularly necessary given the increasing role of stakeholders in overall project identification, design, appraisal and implementation. There is the potential for confusion, 'stakeholder fatigue', and unnecessary expenditure if this issue is not carefully considered. Separate and unrelated stakeholder involvement processes, for the same project proposal, will ultimately be detrimental to effective project planning and implementation. The need to avoid this potential problem strengthens the case for the 'total' integration of EA in project identification and subsequent stages in the project life-cycle.

The use of EA at the project level is being accompanied, slowly but surely, by the use of Strategic Environmental Assessment (SEA) for proposed initiatives such as policy formulation and plan-making. CPP plays a similar role in SEA, but there are sometimes issues of confidentiality (e.g. in policy-making) which mean that CPP may be restricted to intra-government consultation and the input from NGOs is confined to non-public fora.

9.5   Future Trends in CPP

The trend towards the greater use of participatory approaches is, perhaps, the most important trend in CPP. At present, these approaches are more discussed and written about than practised. Nevertheless, it is possible to provide some tentative observations and comments on their implications.

Participatory approaches imply collaboration in, or control over, the EA process. Most proponents and governments are unaccustomed to this way of working and conflicts inevitably will occur. Disagreements occur now, but proponents and governments have many opportunities to ignore them unless they are effectively channelled into local political processes or direct action. The increased likelihood of conflict means that searches for agreement between the stakeholders will be vitally important and mechanisms such as mediation which aim at reaching consensus will be used increasingly.

Mediation, as currently practised, is a voluntary, collaborative process involving face-to-face dialogue and negotiation between representatives of stakeholders. The representatives are accountable to their 'constituents' and attempt to reach agreed solutions. The process is assisted by a specialist mediator who must be independent and acceptable to all stakeholders and remain impartial throughout the process. The mediator is a facilitator not an arbitrator. No stakeholder is forced to accept an agreement and so there is no guarantee that consensus on all issues will be attained. Experience has shown mediation can be successful in certain conditions when areas of disagreement are discrete, well defined and not open-ended. Fundamental value or moral differences cannot be resolved by mediation.

Mediation, as described above, is a product of high income countries and its applicability in other socio-political and cultural settings is not yet known. Mediation assumes, also, that it is possible for all stakeholders to agree the selection of representatives who understand the role of a 'representative' in such decision-making actions. More fundamentally, it may be necessary to limit the number of stakeholders to make mediation manageable. Thus, the selection of key stakeholders, and their representatives, is a major issue – which stakeholders/representatives should be selected and using which criteria? Despite potential difficulties it is certainly worth attempting to use the current mediation model and test its effectiveness before amending or replacing it with an alternative approach.

The use of mediation is most likely in localities with cultural and socio-economic diversity where multiple interests exist. In other situations, for example, rural areas with considerable cultural homogeneity, the diversity of interests will be less and traditional decision-making fora such as the *khotla* in Botswana may be able to reach and present a consensus position. This may be all that is required if the EA is under community control or if the EA is a collaborative exercise. In this situation an agreed community position is one input into a wider decision-making framework which is likely to be more limited in terms of stakeholder numbers than in socially more complex settings.

An example of this situation occurred in the EA for a proposed alumina plant at Weipa in Australia (Newbold 1988). The Aboriginal community expressed a desire to be represented on the steering committee for the social impact assessment component of the EA and the proponent agreed. The local community was content to be represented through its elders. However, the stakeholder policies of external financing agencies will also need to be taken into account – for example, by ensuring that the interests of women have been effectively incorporated into any consensus position adopted by a traditional forum. Such additional requirements will require careful handling.

More questions and difficulties have been raised than answers provided concerning the implementation of more participatory approaches. This reflects insufficient knowledge and experience of managing participatory decision-making processes within both the EA process and the planning and project cycle. However, an awareness of potential difficulties in advance of action provides an early warning which may help to avoid potential problems in the future.

In situations where an EA is under local stake-holder control, participation merges with EA activities and the two become virtually indistinguishable. Examples of this type of EA are very rare, although some interesting cases from countries such as Australia, Canada and the USA, which have a long history of EA and substantial and increasingly vocal populations of indigenous and culturally distinct peoples, may point the way to future EA practice. It may not be too long before the concept of pluralistic EAs is debated in the literature and, eventually, practised in specific contexts (Mulvihill and Jacobs 1998).

The need to ensure effective input from local peoples and the trend toward using traditional ecological knowledge in EA may fuse in the production of EA reports which present multiple impact scenarios. One of these may be based on the application of the 'western' scientific model. However, it will need to be accompanied by other impact scenarios based on 'traditional' science and knowledge. These different versions of the same 'reality' may assist all stakeholders to understand better the range of issues and concerns and, thus, contribute to public/political debate and eventual decision-making. At present, much discussion during EA preparation and at the EA Report stage is undertaken on the basis of a world view not always shared by all participants, thus placing certain of them at a disadvantage.

It is possible to consider this participatory, multi-scenario approach as a viable option for extensive EAs of large-scale, resource-based projects in countries such as Canada where, compared with other countries, there are numerous members of indigenous communities with educational attainments enabling them to play a full role in such EAs. Unfortunately, this situation does not pertain among most indigenous peoples, and indeed most rural populations, throughout huge areas of Africa, Latin America and Asia. Attempting participatory and pluralistic EAs may not be feasible in the short to medium term for both practical and financial reasons. It will be a major challenge, for EA practitioners, to prove this pessimistic prognosis to be wrong.

Discussion Questions

1. How can the increased interest in stakeholder involvement in EA be explained?
2. What are the main difficulties in implementing stakeholder involvement in many low and middle income countries? What techniques/approaches can be used to overcome these difficulties?
3. What are the main stages in EA implementation when stakeholder involvement can make an effective contribution to EA quality?
4. What are the implications for EA practice of an increase in stakeholder *participation*?

Further Reading

There are not many brief overviews of CPP, focusing on developing countries or countries in transition, that are easily available. Although it is not specifically aimed at these countries, a good general source of ideas and information is provided by Roberts (1995). The World Bank Updates (World Bank 1993; 1999) contain useful guidance, but from the perspective of a large multi-lateral financing institution. There is little easily available discussion on CPP that has been produced by EA specialists from developing countries – two exceptions are Kakonge (1996) and Mutemba (1996).

References

Davis, SH and Soeftestad, LT (1995) *Participation and Indigenous Peoples*, Environment Department Paper No. 021, Environment Department, World Bank, Washington DC

Kakonge, JO (1996) Problems with public participation in EIA process: examples from sub-Saharan Africa, *Impact Assessment* **14**: 309–320

Mulvihill, PR and Jacobs, P (1998) Using scoping as a design process to promote appropriate and relevant environmental assessment, *Environmental Impact Assessment Review* **18**: 351–369

Mutemba, S (1996) Public participation in environmental assessments for bank-supported projects in Sub-Saharan Africa, in *Environmental Assessment (EA) in Africa: A World Bank Commitment (Proceedings of the Durban, World Bank Workshop June 25, 1995)* Goodland, R *et al.* (eds), Environment Department, World Bank, Washington DC

Newbold, L (1988) Weipa: Managing the Impact of a Major Mining Project on the Physical and Social Environment of a Remote Area, Unpublished Conference Paper

OECD Development Assistance Committee (1994) *Capacity Development in the Environment: Toward a Framework for Donor Involvement*, Background Paper prepared for the OECD/DAC Task Force on Capacity Development in the Environment OECD Development Assistance Committee, Paris

Pretty, JN (1993) *Environmental Assessment and Participation: Some Challenges and Dangers* IIED, London

Pretty, JN *et al.* (1993) *A User's Guide to Participatory Inquiry* IIED, London

Roberts, R (1995) Public involvement: from consultation to participation, in *Environmental and Social Impact Assessment* Vanclay, F and Bronstein, D. (eds), Wiley, Chichester, pp. 221–246

United Nations Environment Programme (UNEP) (1996) *Environmental Impact Assessment: Issues, Trends and Practice* UNEP, Nairobi

World Bank (1993) *Public Involvement in Environmental Assessment: Requirements, Opportunities and Issues*, Environmental Assessment Sourcebook Update Number 5, Environment Department, World Bank, Washington DC

World Bank (1999) *Public Consultation in the EA Process: A Strategic Approach*, Environmental Assessment Sourcebook Update Number 26, Environment Department, World Bank, Washington DC

10

Integrating Appraisals and Decision-making

Norman Lee

10.1 Introduction

The main theme of this chapter is the integration of environmental assessments (EAs) into decision-making relating to projects in developing countries. It is particularly concerned with the strengths and weaknesses of linkages between the EIA process and the project cycle.

- Section 10.2 clarifies certain conceptual and contextual issues relating to linkages between appraisal and decision-making
- Section 10.3 reviews the main deficiencies of current practice in integrating EIAs into decision-making within the project cycle
- Section 10.4 reviews different methodological approaches to integrated appraisals for use in decision-making
- Section 10.5 summarizes the main conclusions and makes recommendations to improve current practices.

A more extended analysis of the relationship between appraisal, decision-making and sustainable development in developing countries is contained in Lee and Kirkpatrick (1999).

Environmental Assessment in Developing and Transitional Countries. Edited by N. Lee and C. George.
© 2000 John Wiley & Sons, Ltd.

10.2   Conceptual and Contextual Issues

Types of Decisions and Their Timing

Typically, EIA regulations specify that the findings of the environmental impact statement (EIS) and the results of consultations based upon it shall be 'taken into consideration' in reaching a decision on the authorization of a project or granting an environmental permit. This might imply that the EIA process is only linked to one decision point in the project cycle. However, good international practice suggests that the environmental assessment process should proceed in tandem with the project cycle and link into a number of decision points at different stages in that cycle (that is, whilst the project is being *planned*, *approved* and *implemented*). These decisions involve different kinds of decision-makers and other stakeholders (e.g. developer, regulatory or planning authority, aid agency) who function in different decision-making contexts.

Differences in Decision-making Contexts

Decisions are not made in a vacuum. They are influenced by:

- Who is involved in taking the decisions and their motivations
- The social, political and economic circumstances and the regulatory, procedural and institutional constraints within which the decisions are taken

Taken together, these comprise the decision-making context.

In the case of many projects there will be different decision-makers involved at different stages in the project cycle and their motivations and decision criteria will also differ. A developer in the private sector may be motivated by profit and only be interested in complying with minimum environmental requirements at least cost to himself. The competent authority, which decides whether or not to authorize a project, will consider its statutory objectives. If it has a statutory responsibility to protect certain environmental resources (rather than all types of environmental resources) its decision criteria may be confined to these. In other cases, an authority may also be required to take other kinds of material considerations into account and its decision criteria will be broadened to reflect this. If the authority's main responsibilities lie in another sector (e.g. the transport or energy sector) environmental considerations may be subordinated to its sectoral objectives.

One important consequence of the above situation is that both appraisal and decision-making are prone to conflicts of interest between different stakeholders which have different goals and decision criteria. Hence, conflict avoidance and conflict resolution become an integral part of the appraisal and decision-making process and help to shape the appraisal and decision-making criteria which are finally applied.

These potential conflicts are handled in different ways in different decision-making contexts. It is possible to distinguish three very different approaches:

- The dictatorial approach – one dominant decision-maker makes his/her own decision according to his/her own criteria, with little reference to either professional analysis or the opinions of others
- The professional analysis approach – the decision-maker plays a passive rôle, relying upon the advice of his professional advisors which is based on clearly defined appraisal procedures and decision rules.
- The consultative approach – the decision-maker plays a passive rôle and the professional's rôle is confined to being a facilitator of the consultative process. The principal stakeholders and interest groups are involved in identifying and negotiating a mutually acceptable outcome which the decision-maker can then endorse.

In the real world, hybrids of two, or even all three, of these approaches are found. However, the balance between them varies greatly, both within and between individual countries.

How does the environmental assessment process fit into these different decision-making contexts? In principle, it contains elements of all three approaches. It has a strong technical element (reflected in the technical studies which underpin the whole process and which are documented in the EIS or SEA report). It has a consultative element (reflected in requirements for consultation and public participation) and it acknowledges the ultimate authority of the decision-maker in approving, modifying or rejecting the proposed action. However, if the mix of these three elements in the EIA process is at variance with the mix in the decision-making context to which it is linked, then:

- Conflicts may occur and opposition to the EA process will intensify, rendering it ineffectual
- Non-compliance problems may arise which also make it ineffectual

Experience in those developing countries and countries in transition which have widely different decision-making contexts to those in the developed economies indicate that wholesale adoption, without significant modification, of the western EIA model can have disappointing consequences.

10.3 Current Practice

This review of current practice relates to the integration of environmental assessment at different stages of the project cycle in developing countries. It focuses on the following weaknesses, many of which also occur in other types of country:

- Starting the EIA process too late in the project cycle
- Lack of co-ordination between EIA and project planning
- Limitations within EISs
- Deficiencies in consultation arrangements

- Inadequate linkages between EISs, other appraisals, consultations and decision-making
- Weak links between the EIA process and project implementation

Proposals relating to improvements in current practice are contained in Section 10.5.

Starting the EIA Process too Late

A common deficiency is a failure to integrate EIA into decision-making at early stages in the project cycle because the EIA process itself has started too late (Sadler 1996). Evidence on the late starting of the EIA process in India, Malaysia and Tanzania has already been presented in Chapter 8 (Section 8.5). Additional evidence, relating to Indonesia, the Philippines and Sri Lanka, is shown in Box 10.1.

Box 10.1   EIA – Project Planning Linkages in Sri Lanka, Philippines and Indonesia

'EIA tends to be seen as a single-shot regulatory activity that results in the imposition of environmental strictures, rather than as an integral part of all phases of national and project-level planning and design and as a mechanism for helping decision-makers balance competing economic, technical, social and environmental concerns.'

Sri Lanka 'As far as national economic and local development planning and implementation processes are concerned, the EIA process fits very poorly as of now. There appears to be great resistance within planning agencies to use it as a planning tool and it can hardly be said to be integrated into their processes at all.'

Philippines 'Recent experiences encountered in the implementation of major projects in the country show that, despite the conduct of EIAs, serious difficulties are still faced by project implementers because environmental consequences are not sufficiently considered in the early stages of the project cycle. As a result, the entry point of environmental assessment in the project cycle is not early enough to be useful in decision-making and EIA is oftentimes seen as merely another bureaucratic requirement to hurdle.'

Indonesia 'The link between AMDAL [the EIA process] and the licensing system becomes dysfunctional when the proponent reaches the stage of asking for an operational licence . . .. AMDAL can only be effective if it is completed before permits and licences are issued.'

'Better and more explicit linkages are needed between the planning, EIA review, permitting, compliance monitoring and enforcement stages.'

Source: Smith and van der Wansem (1995) pp. 19–22 (extracts).

Lack of Co-ordination Between EIA and Project Planning

Here, also, the evidence of deficiencies is quite extensive. For example, the World Bank reports:

> Many EAs are still not an integral part of project preparation. Rather, they tend to be relatively independent assessments, usually carried out by consultants that may be

detached from the broader project preparation process. Often this approach is ingrained in borrower EA procedures and cannot easily be adjusted . . .. Another problem is that borrowers are frequently interested in cutting costs, which may lead to the utilisation of consultants who do not have sufficient expertise to take an active part in the project design process. (World Bank 1997, pp. 51–52).

Other relevant factors that are mentioned are limited institutional capacity, the low status of EIA as a planning and decision-making tool, and the limited authority of the environmental agencies. Similar points are made in relation to India, Malaysia and Tanzania in Section 8.5 and, in relation to Sri Lanka, Philippines and Indonesia in Box 10.1. Another illustration, relating to the environmental assessment of a project in Tanzania, is provided in Box 10.2.

Box 10.2   A Graphite Mining Project in Tanzania

The environmental assessment process started formally with the feasibility study which included a brief discussion of environmental issues. However, the African Development Bank's financing procedures obliged the mining company to prepare a more adequate EIS. AfDB prepared terms of reference for the EIA and the company sub-contracted its preparation, both taking place after the design of the plant and mining operation had been finalized.

The environmental impact study was undertaken principally as a 'stand-alone' exercise. The proponents provided the background information required for the consultants to prepare the EIS, but there was no further integration between the EIA process and project design. The EIS focused tightly on the design prepared during the feasibility study, and did not explore alternative options for design, plant siting and mine waste disposal plans. In a more general sense, the EIS would appear to have had no impact on the siting, design and operation of the project.

Source: Mwalyosi and Hughes (1998) pp. 51–52.

Limitations Within EISs

Three main types of limitations can be distinguished in the practical usefulness of EISs for decision-making purposes:

- *Timing.* The EIS may not be completed and made available at a sufficiently early stage for consultation and project authorization purposes. Ibrahim (1992) and Rout (1994) provide relevant information for Malaysia and India but this is also a significant problem in many other countries
- *Content and methodology.* The form of the analysis used, and the content of the findings in the EIS, may not be most appropriate for decision-making purposes. 'Analysts often appear to have no clear idea as to what type of information is needed by the decision-makers.' (Biswas and Agarwala 1992, p. 240)
- *Presentation.* The findings of the environmental assessment may be presented in a form which is not sufficiently useful for consultation and decision-making purposes. The problems are of various kinds: inadequate or non-existing, non-technical summaries; too much data and too little analysis and interpretation;

unnecessary complexity; bias; insufficiently clear and precise recommendations and mitigatory proposals (see Section 8.4 for further details)

Deficiencies in Consultation Arrangements

A further deficiency relates to inadequacies in consultation and public participation which restrict its effectiveness both within the EA process and in links to decision-making in the project cycle. The following weaknesses have been observed (see also Box 10.3).

- *Timing.* Formal requirements for consultation and public participation (CPP), where they exist, are often confined to the later stages of the EIA process, after the EIS has been completed. This weakens the performance of the pre-EIS submission phases and the linkages with planning and design decisions during earlier stages in the project cycle
- *Inadequate regulations and consultation practice.* Even where regulatory provision exists for consultation and public participation, for example, after EIS submission, its practical effectiveness may be limited by: inadequate awareness of, and publicity for, CPP opportunities; restrictions on the availability of EISs; insufficient time in which to make comments; capacity and other resource constraints on environmental agencies, NGOs and the general public in organizing and presenting their views; inadequacies in processing CPP findings and presenting them in a form which is useful for decision purposes.

Box 10.3   World Bank Experience in CPP Within the EIA Process and Project Cycle

'Consultation continues to be weak in many projects, especially outside the context of involuntary settlement . . .. Progress also appears to have been stronger in terms of consultation with local NGOs than with affected community representatives. In too many cases, women and the poor are not reached. Some EAs still do not sufficiently document the consultation process and results even where the process has been strong and the results positive . . .. Bank-wide, less than half of the category A projects reviewed had any record of consultation at the draft EA report stage, and NGOs and specialists were involved more often than local communities . . .. The Bank needs to ensure that the EA consultant team adequately understands the consultation requirements [of the Bank] and – where it is responsible for the consultation process – that it has sufficient experience in carrying out and documenting meaningful consultation' (World Bank 1996, pp. 34–40)

Inadequate Linkages Between Appraisals, Consultations and Decision-making

A fifth deficiency is associated with the inadequate, combined use of the EIS, other forms of appraisal and consultation findings in reaching decisions relating to the approval or financing of the project. This may arise from a number of causes:

- *Timing.* If the appraisal and consultation findings are not available when key decisions are being taken, they will have little or no influence

- *Inadequate integration with other appraisals.* Decisions, explicitly or implicitly, are often based upon economic, social and broader planning considerations, as well as environmental considerations. However:
  - The methodologies used to produce integrated appraisals may be inadequate or inappropriate in the decision-making context in which they are applied
  - The Report which contains integrated findings, and which is used for decision-making purposes, may not be sufficiently clear, balanced and technically sound for this purpose
- *Transparency of decision-making.* The accountability of decision-makers, for taking EIS and consultation findings into consideration, depends greatly upon the visibility of their decisions and of the reasoning upon which they are based. However:
  - The decisions, including any environmentally relevant conditions attached to them, may not be made public
  - The decisions and their accompanying environmental conditions may not be expressed in sufficiently detailed and precise terms to provide a satisfactory basis for their subsequent enforcement
  - The reasons for the decisions, taking into consideration the contents of the supporting appraisal documentation, may not be recorded or made public

A number of these deficiencies are frequently observed, as illustrated in Box 10.4.

Box 10.4   Examples of Weak Links Between EIA Findings and Decision-making

'Another issue that arose in all case studies [relating to the Philippines, Indonesia and Sri Lanka] is the lack of links between EIA findings and project permit and licensing . . .. In all three countries, the environmental review must be completed before construction and operating permits and licences are granted . . .. Yet success in achieving this objective has been mixed.' (Smith and van der Wansem 1995, p. 21)

'. . . of the 26 EIAs reviewed in this study, only three have tangibly influenced decision-making, and two of these may yet proceed in much their original form' (Mwalyosi and Hughes 1998, p. 79)

'Once an EA report has been received and cleared by the Bank, the next step is to formally reflect its main findings in the project documentation, i.e. in the Staff Appraisal Report and legal documents. This is a critical point in the EA process in terms of its effectiveness visa vis final project design. Performance has generally improved. However, the references [to environmental plans and recommendations derived from the EA in legal documents] are usually general in nature and rarely mention specific actions or criteria to be met' (World Bank 1996, pp. 55–56)

Weak Links Between the EA Process and Project Implementation

The links between the EA process and the later (post-authorization) stages of the project cycle are often weak (Goodland and Mercier 1999) due to:

- *Deficiencies in environmental management plans (EMPs).* EMPs, which indicate how projects are to be implemented in accordance with well-defined environmental criteria, are insufficiently prepared during the EA process and then used when formulating conditions in authorizations and permits

- *Deficiencies in monitoring compliance and in enforcing compliance, through the use of legal instruments and financial penalties.* Both of these kinds of shortcomings are documented in various studies (see Box 10.5) and are examined in greater detail in Chapter 11.

Box 10.5   Environmental Assessment and the Post-authorization Phase in the Project Cycle

'The use of environmental impact monitoring and management plans in implementing the findings and recommendations of EIA studies is a key issue. Often, these plans are neglected or not effectively enforced.' (Smith and van der Wansem 1995, p. 21)

'EIA, as it is practised now, ends immediately after the environmental clearance of a project has been received. Compliance monitoring is seldom carried out, either by the project authorities or by the responsible government agencies' (Biswas and Agarwala 1992, p. 240).

'Despite . . . advances, monitoring and EIA follow-up mechanisms remain poorly developed . . .. World-wide, the frequency with which follow-up is either absent or perfunctory amounts to a systematic weakness of the EIA process' (Sadler 1996, p. 126)

'In most of the case studies reviewed, the EIA process ended with the submission of the EIS. In no case did EIA practitioner involvement continue during the implementation or post-completion stages of the project. Post-completion monitoring seems to be particularly poor in this respect.' (Mwalyosi and Hughes 1998, p. 69)

The deficiencies which have been identified in linkages between environmental assessment and decision-making at the project level can be grouped into the following categories: regulatory, procedural, methodological and institutional. Similar kinds of deficiencies occur in integrating EA findings into decisions relating to the formulation, approval and implementation of policies, plans and programmes. Indeed, because SEA regulations, procedures, assessment methodologies and institutional capabilities are less developed than for project level EIA, the deficiencies can be considerably greater, even though they may be less immediately visible.

10.4   Appraisal Methodologies

The choice of the most appropriate appraisal methodology depends upon the decision-making context and the objectives of those involved in making the decision. A number of different kinds of appraisal methods are reviewed below. In each case (with the partial exception of the first) the focus is on appraisal and decision-making situations where environmental impacts are not the only consideration in reaching decisions.

Compliance with Emission and Other Environmental Standards

In some countries (for example, in Germany, Russia and certain other countries in transition), a developer may be granted an environmental licence or approval if his projected discharges can be shown to comply with specified standards. In this case, the appraisal should predict the likely size of the environmental impacts associated with the development and determine whether these comply or not with the standards. If the EIA process is linked to such a system, then its EIS (or equivalent) should record this information in a suitable form for the appraising authority to use. The developer, for his part, and assuming he is a profit maximizer, will wish to know that his project specification is the least cost way of achieving the required standards.

In practice, the appraisal criteria may be less straightforward than these. The developer may be a profit maximizer who is prepared to ignore the standards if the fines he would have to pay for non-compliance are less than the costs of pollution abatement to keep within the standards. Standards may not have been established for all of the discharges. Also, some of the standards may be very old and there may be a strong public feeling that they are either too strict or not strict enough. Crucially, the licensing authority may have some discretion as to whether to grant the environmental licence or not, having regard to wider considerations than compliance with environmental standards. In these circumstances, the simple environmental appraisal described in the previous paragraph is incomplete and not fully relevant for decision purposes by both developer and competent authority. Nor is it necessarily sufficient for consultation and public participation purposes.

Goals Achievement Analysis

An extension to the above methodology is to appraise proposed developments and to compare alternatives according to the degree of their compliance with all of their goals and not only with environmental standards. This may be undertaken using goals achievement matrices (Hill 1968). This requires that all relevant goals are clearly specified in an operational form and that the appraisal (of which the environmental assessment is a part) contains predictions which can be used to check the likely achievement of all goals. If one alternative will meet all of the goals and none of the other alternatives will do so, the choice between them is straightforward. If more than one (or none of them) will do so, then additional considerations, involving some trade-offs (see Multi-criteria Analysis below), are likely to be needed.

Cost-benefit Analysis

The main decision criterion of some public authorities and government agencies is that any new development should give rise to greater social benefits than social costs. In these cases, the appraisal might use some form of social cost-benefit analysis (CBA). This should include all benefits and costs to society – economic, social and environmental. The standard methodology is to quantify all benefits and costs over the expected life of the development, to evaluate them using monetary units and to convert them to a common (base) point in time using discounting procedures. The benefits and costs should then be compared and, if the discounted social benefits exceed the discounted social costs (i.e. the net present values [NPV] is positive), it is concluded that the development is likely to be in the public interest. If investment capital is in short supply, development projects may be prioritized according to the predicted size of the NPV of each project relative to its capital investment requirements (Curry and Weiss 1993; Kirkpatrick and Weiss 1996).

In such situations, the environmental assessment may be used as an input to the cost benefit analysis (World Bank, 1998). Ideally it should contain predictions of the magnitude of the environmental impacts (negative and positive) which, using economic valuation methods as described in Chapter 6, are then converted into monetary estimates for inclusion with other economic and social estimates in an overall social CBA. This is illustrated in Box 10.6.

Box 10.6   Cost-benefit Analysis of a Forest Development Project in Nepal

This is a project in a hilly area which could bring benefits through reduced soil erosion, increased productivity of land within the watershed plus increased fuel wood and fodder for the local community (i.e. a mixture of economic, social and environmental benefits). If this does not proceed, then it is expected that the conversion of forest to arable will continue. Which is preferable, having regard to the costs and benefits of the forest development project in comparison with a baseline of continued conversion to arable? The CBA estimated that the discounted benefits of the project would be greater than the discounted social costs if the discount rate was less than 8.5% per annum.

Source: Dixon, JA *et al.* 1994, p. 120 *et seq.*

CBA is a potentially powerful tool of integrated appraisal where an overall comparison between benefits and costs to society as a whole is required. It has influential supporters among banking and other financial institutions which are accustomed to use economic appraisals as an aid to their decision-making. However, there are a number of controversies relating to its use which are briefly reviewed below.

Technical issues

The technical issue which generates most concern is the extent to which environmental impacts can be satisfactorily valued in economic terms for inclusion in a CBA (see Chapter 6 where this issue is examined in some detail). Other technical sources of concern in particular cases are:

● Has a sufficient range of alternatives been investigated
● Is the scope of the benefits and costs which have been assessed sufficiently comprehensive
● Has the degree of uncertainty been sufficiently taken into account in the prediction and valuation of costs and benefits

- Have appropriate discount rates been used in the calculation of net present values
- Is the distribution of costs and benefits between different sectors of society sufficiently reflected in the appraisal findings?

Often the main concern is not over the CBA methodology in general but its practical application in particular cases. A number of similar kinds of technical issues can arise in the use of environmental and social assessment methodologies (Lee and Kirkpatrick 1997a,b).

Consultation and transparency issues

Part of the concern over the use of CBA relates to its perceived lack of transparency and limited public involvement in the appraisal. Whether this is inherent to the methodology or simply ingrained in practice is an issue deserving further examination. As greater use is made of interview and survey techniques in certain valuation studies (such as contingent valuation studies; Common 1999) and becomes more closely associated with environmental and social assessments and their participative cultures, CBA practice may itself become more open and participative.

Multi-criteria Analysis

Where CBA is considered inappropriate, some form of multi-criteria analysis (MCA) may be used instead. This avoids converting all costs and benefits into monetary units and appraising projects according to a single criterion. MCA exists in a variety of forms which differ considerably in their sophistication and complexity as well as in their information requirements (Nijkamp *et al.* 1990; DETR 1998; Lichfield 1996). The better known methods include:

- *Trade-off analyses.* These are most useful where there are a limited number of alternatives and impacts to be compared
- *Pair-wise comparisons.* These are more sophisticated ways of making comparisons where the numbers of alternatives and impacts are greater
- *Goals-achievement matrices.* These, as mentioned earlier, are used when comparing alternatives according to their likely attainment of a number of different objectives
- *Planning balance sheets.* These show the distribution of impacts between the different community groups and stakeholders who will be affected by each of the alternatives
- *Scoring and weighting systems.* These compare alternatives by standardizing the scores of different impacts on a common scale and weighting each type of impact according to its assumed relative importance

In general, it is the simpler methods which are more widely used, particularly when presenting the results of appraisals to decision-makers and for public use.

Additionally, there is a growing number of cases where CBA are being used within the broader framework of MCAs (DETR, 1998). Also, there is some tendency towards greater transparency in MCA studies and towards the greater involvement of stakeholders in the appraisal process (Voogd 1997).

Expert Judgement

Faced by the shortcomings and, in certain cases, complexities of the different appraisal methodologies, it is not surprising that the most common method of appraisal used, in practice, by decision-makers is expert judgement (i.e. the judgement of an expert adviser or the use of their own judgement). However, like all other methods, it should be submitted to critical assessment:

- What appraisal criteria and data are the experts using?
- What consultations are they undertaking?
- What analyses are they carrying out and what are the value judgements and other assumptions upon which they are based?
- What are the professional qualifications and experience of the experts making such judgements or appraisals?
- How are the expert judgements being presented to decision-makers and stakeholders and how far are they being substantiated?

The more transparent and explicit is the process of expert judgement, the more closely it will be possible to evaluate it and compare it with the more formal appraisal methodologies outlined previously.

10.5   Conclusions

The performance of the environmental assessment process is dependent on the degree of success in integrating assessment findings into decision-making in the planning and project cycle. The linkages between appraisal and decision-making are not well researched and understood, particularly in less developed countries and countries in transition. Such evidence as exists, however, indicates that they are often deficient. The chapter concludes with a checklist of suggestions for improving the situation, based upon the earlier findings.

1. Identify the key decisions in the planning and project cycle into which environmental assessments should be integrated.
2. Identify when these decisions are to be taken, who/which organization makes these decisions and the goals of those responsible for making them.
3. Be aware of the social, political and economic situation in which appraisal and decision-making takes place. Also be aware that the strengthening of appraisal procedures and methods can help to modify and improve the effectiveness of development planning and decision-making practices.

4. Weak (or non-existent) *procedural* linkages between environmental assessment decision-making may be strengthened by:
   - Starting the EA process earlier in the planning and project cycle
   - Strengthening arrangements for co-ordination between environmental assessment and policy, plan and project preparation
   - Remedying the deficiencies in EISs (and in EA documentation relating to policies, plans and programmes) concerning the timing of their completion, their content and methodology and the presentation of their findings
   - Strengthening consultation and decision-making linkages through better timing of consultations, more specific consultation requirements and better consultation practice
   - Strengthening the integration of environmental assessments with other social and economic appraisals and improving the transparency of the decision-making process itself
   - Strengthening the linkage between the EA process and plan and project implementation through more effective use of environmental management plans, monitoring compliance and enforcing compliance
5. *Appraisal methodologies* should be selected which are more closely related to the decision-making context in which they are to be used, and the goals of the bodies involved in making the decisions. They should be user-friendly, feasible in terms of data requirements and available appraisal expertise, and sufficiently transparent. They should also address any potential inconsistencies between separately undertaken assessments of environmental, social and economic impacts.

The agenda for strengthening the integration of appraisal and decision-making is ambitious. To achieve the fullest benefits, changes within both the planning and project cycle and in the decision-making context will be needed as well as changes in appraisal methods and procedures. In many situations, this may only be achieved, on a step-by-step basis, over a number of years.

Discussion Questions

1. In what ways and to what extent should the decision-making context shape the methods and procedures used in the appraisal of policies, plans and programmes or projects?
2. What are the main institutional and procedural obstacles to the effective integration of environmental assessments into decision-making in developing countries and countries in transition? How might these obstacles be overcome or significantly reduced?
3. Compare any *two* methodologies which may be used to integrate environmental assessments into overall appraisal and decision-making, indicating their principal advantages and disadvantages in a developing country or country in transition context.

Further Reading

A further analysis of issues relating to integrated appraisal, decision-making and sustainable development is contained in Lee and Kirkpatrick (1999, ch. 1). Additional information and analysis on the relevance of the decision-making context to the choice of appraisal methods and procedures can be found in Hilden *et al.* (1995), Voogd (1997) and UNEP (1996). Further information on practice and shortcomings in procedural and institutional linkages between environmental assessment, overall appraisal and decision-making can be found in Biswas and Agarwala (1992), Ministry of Housing, Spatial Planning and Environment (1994), Smith and van der Wansem (1995), Mwalyosi and Hughes (1998) and Dalal-Clayton and Sadler (1998). Empirical studies relating to the relationship between public participation and decision-making in countries in transition can be found in Regional Environmental Centre (1998). Proposals relating to good practice in integrating environmental assessment into overall appraisal and decision-making for aid agencies are contained in OECD (1996). Reviews of methodologies for linking environmental assessment to integrated appraisals and decision-making can be found in Canter (1996), Canter and Sadler (1997), Dixon *et al.* (1994), DETR (1998), World Bank (1998), Winpenny (1995), Curry and Weiss (1993), Kirkpatrick and Weiss (1996), Hill (1968), Lichfield (1996), Nijkamp *et al.* (1990) and Voogd (1997). Most of the above literature relates to project level appraisal and, with the exception of the cost-benefit literature, the literature concerning appraisal and decision-making in developing countries and countries in transition is relatively limited.

References

Biswas, AK and Agarwala, SBC (1992) *Environmental Impact Assessment for Developing Countries*, Butterworth-Heinemann, Oxford

Canter, LW (1996) *Environmental Impact Assessment* 2nd edn, McGraw Hill, New York (chapters 15 and 16)

Canter, L and Sadler, B (1997) *A Tool Kit for Effective EIA Practice – Review of Methods and Perspectives on their Application*, University of Oklahoma, Norman, OK

Common, M (1999) Environmental cost benefit analysis and sustainability, in *Integrated Appraisal and Sustainable Development in a Developing World*, Lee, N and Kirkpatrick, C (eds), Edward Elgar, Cheltenham (in press)

Curry, S and Weiss, J (1993) *Project Analysis in Developing Countries*, Macmillan, London

Dalal-Clayton, B and Sadler, B (1998) *The Application of Strategic Environmental Assessment in Developing Countries*, International Institute for Environment and Development, London

Department of the Environment, Transport and the Regions (DETR) (1998) *Review of Technical Guidance on Environmental Appraisal* DETR, London

Dixon, JA, Scura, LF, Carpenter, RA and Sherman, PB (1994) *Economic Analysis of Environmental Impacts*, Earthscan, London

Goodland, R and Mercier, JR (1999) The evolution of environmental assessment in the World Bank: from 'approval' to results, *Environment Department Papers No 67*, World Bank, Washington DC

Hilden, M *et al.* (1998) *EIA and its Application for Policies, Plans And Programmes in Sweden, Finland, Iceland and Norway*, Nordic Council of Ministers, Copenhagen

Hill, M (1968) A goals-achievement matrix for evaluating alternative plans, *Journal of the American Institute of Planners* **34**: 19–28

Ibrahim, AKC (1992) An Analysis of Quality Control in the Malaysian Environmental Impact Assessment Process, unpublished MSc dissertation, University of Manchester, Manchester

Kirkpatrick, C and Weiss, J (eds) (1996) *Cost-benefit Analysis and Project Appraisal in Developing Countries*, Edward Elgar, Cheltenham

Lee, N and Kirkpatrick, C (1997a) Integrating environmental assessment with other forms of appraisal in the development process, in *Sustainable Development in a Developing World: Integrating Socio-economic Appraisal and Environmental Assessment*, Kirkpatrick, C and Lee, N (eds), Edward Elgar, Cheltenham, pp. 1–24

Lee, N and Kirpatrick, C (1997b) The relevance and consistency of EIA and CBA in project appraisal, in *Sustainable Development in a Developing World: Integrating Socio-economic Appraisal and Environmental Assessment*, Kirkpatrick, C and Lee, N (eds), Edward Elgar, Cheltenham, pp. 125–138

Lee, N and Kirkpatrick, C (eds) (1999) *Sustainable Development and Integrated Appraisal in a Developing World*, Edward Elgar, Cheltenham (in press)

Lichfield, N (1996) *Community Impact Evaluation*, UCL Press, London

Ministry of Housing, Spatial Planning and the Environment (1994) *Use and Effectiveness of Environmental Impact Assessment in Decision-Making* Report No 49, VROM, The Netherlands

Mwalyosi, R and Hughes, R (1998) *The Performance of EIA in Tanzania: an Assessment*, International Institute for Environment and Development, London

Nijkamp, P, Wietveld, P and Voogd, H (1990) *Multi-criteria Evaluation in Physical Planning*, North Holland, Amsterdam

OECD (1996) *Coherence in Environmental Assessment: Practical Guidance on Development Co-operation Projects*, OECD, Paris

Regional Environmental Center (1998) *Doors to Democracy: Current Trends and Practices in Public Participation in Environmental Decision-Making*, vol 1 Central and Eastern Europe, vol 2 Newly Independent States, REC, Szentendre, Hungary

Rout, DK (1994) An Analysis of the EIA Process and EIA Reports produced for selected industrial developments in the State of Orissa in India, unpublished MSc dissertation, University of Manchester, Manchester

Sadler, B (1996) *Environmental Assessment in a Changing World: Final Report*, Ministry of Supply and Services, Canada

Smith, DB and van der Wansem, M (1995) *Strengthening EIA Capacity in Asia: Environmental Impact Assessment in the Philippines, Indonesia and Sri Lanka*, World Resources Institute, Washington, DC

UNEP (1996) *EIA Training Resource Manual* UNEP, Nairobi (Topic 10 on Decision-Making, Topic 12 Project Management)

Voogd, H (1997) The changing rôle of evaluation methods in a changing planning environment: some Dutch experience, *European Planning Studies* **5**: 257–266

Winpenny, J (1995) *The Economic Appraisal of Environmental Projects and Policies: A Practical Guide*, OECD, Paris

World Bank (1997) *The Impact of Environmental Assessment: Second Environmental Assessment Review*, World Bank, Washington DC

World Bank (1998) *Economic Analysis and Environmental Assessment*, EA Source Book Update 23, World Bank, Washington DC

11

Environmental Monitoring, Management and Auditing

Clive George

11.1 Introduction

If the road to hell is paved with good intentions, environmental assessments which end at the decision-making stage make costly and misleading paving stones. Their good intentions are likely to come to nothing if they are not monitored.

Effective environmental assessment is a process rather than an isolated event, and is itself part of the broader process of environmental planning and management (Chapter 2). After a decision has been taken to proceed with a project or other development action, the environmental assessment process should continue on into the implementation stage, beyond into actual operation, and ultimately into decommissioning or the next planning or policy-making cycle. During these later stages of the process there is often considerable overlap between the monitoring of EA and other aspects of environmental management. Pollution control, waste management, water resource management, landuse planning and other aspects of development planning and policy-making all involve various forms of monitoring of the environment, and of activities which affect it. The follow-up monitoring of EA cannot be treated in isolation from these other aspects of environmental management.

Arrangements for handling these other aspects may be fairly rudimentary in some low and middle income countries, and in some cases EA may be the only instrument for dealing with them. When overseas funding agencies are involved, their monitoring responsibilities overlap with those of national authorities, adding yet more complexity.

Environmental Assessment in Developing and Transitional Countries. Edited by N. Lee and C. George.
© 2000 John Wiley & Sons, Ltd.

The term *monitoring* is used here in its broadest sense. It embraces all relevant *checks* of activities, impacts and environmental parameters, to ensure that they are in accordance with plans, predictions and approval conditions. This includes various forms of *audits* and *inspections*, as well as regular and ad hoc *observations* and *measurements* of the environmental parameters which may be affected. An *audit* is a check carried out by an independent body (a 'calling to account').

Within the EA process itself, monitoring serves two prime purposes:

- To ensure that the action is implemented as described in the assessment
- To ensure that its impacts are no greater than those predicted in the assessment

Additionally, it can provide valuable feedback for use in future assessments.

While these objectives are fairly straightforward, achieving them can be quite complex. As well as multiple overlaps with other aspects of environmental management, monitoring involves numerous actors in the EA process itself. Investors, developers, operators, NGOs, the public and all the relevant competent authorities all have different interests and responsibilities, all of which have to be co-ordinated, for the process to be fully effective. Some aspects of the developer's or operator's role may be mandatory, while others may be part of good management practice.

This chapter reviews the principles and methods that can contribute to effective monitoring of EA, in relation to other aspects of environmental management, in developing countries and countries in transition. Section 11.2 summarizes the relationships between environmental assessment and the various aspects of monitoring described subsequently. Sections 11.3–11.6 discuss environmental management plans (EMP), environmental management systems (EMS), supervisory monitoring, and a number of types of environmental audit, in the context of individual projects or activities. Section 11.7 reviews the application of similar approaches to the less well established area of strategic assessment of policies, plans and programmes. The final section summarizes key findings and considers future improvements.

11.2 Integrating Environmental Assessment, Environmental Management Plans and Environmental Management Systems

The aim of environmental assessment is to ensure that, before a project or other action is approved, the environmental impacts which it is likely to have during its entire life cycle are understood, and are acceptable. For this aim to be achieved in full, monitoring of the action against the predictions made in the assessment must be carried out throughout the action's life cycle. Provisions must also be made for remedial action in the event that predictions prove to be invalid.

For a development project, the principal stages of the life cycle that need to be checked or monitored are design, implementation, operation and decommissioning. Environmental assessment itself is a check on the design. To be fully effective, the EA needs to include provisions for further checks during the later stages of the cycle. Figure 11.1 shows schematically how this may be achieved, in principle, in a fully integrated manner.

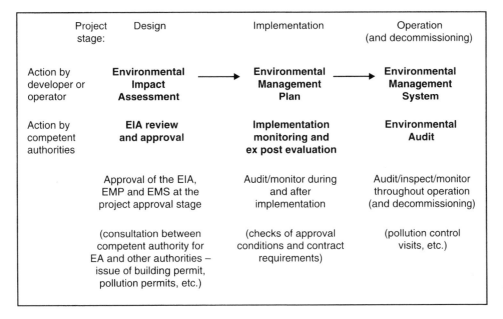

Figure 11.1   *Integrated Environmental Management*

In Figure 11.1 and throughout this chapter, we use the term *Environmental Management Plan* to mean a description of all the relevant actions that will be taken by the developer, including monitoring of impacts and establishing capacity for ongoing management, during the implementation or construction stage of a project, up to and including post-commissioning. The definition used by the World Bank (World Bank 1999) is 'the set of mitigation, monitoring and institutional measures to be taken during implementation and operation to eliminate adverse environmental and social impacts, offset them, or reduce them to acceptable levels'. By implication, an EMP should include some form of *Environmental Management System*, by which is meant a similar description of actions that will be taken by the operator, including monitoring of impacts, during the subsequent operational phase of the project, up to and including decommissioning. *Implementation monitoring* and *ex post evaluation* are forms of monitoring carried out within this framework by competent authorities, as is one particular form of *Environmental Audit*.

At the project approval stage, the competent authority for EA can ensure that the developer makes suitable provisions for appropriate management and monitoring during all the later stages of the project's life cycle, by requiring an EMP to be submitted at the same time as the EIA (or as part of it), and by requiring an EMS to be defined at the same time also (or as part of the EMP). Where this is the case, the competent authority will not approve the project unless the EIA, EMP and EMS proposals are all satisfactory.

A more radical approach that has been suggested (Goodland and Mercier 1999) is that the EMP should be the prime document, with an EMS embedded in it, and with impact prediction relegated to an annex explaining why mitigation measures are necessary. While this would have the advantage of focussing attention on implementation, which is currently a weak area of EA practice, it may dilute attention to environmentally sound design, and understate the need for environmentally sound operational management. Both of these are just as important, and all three are distinct processes, with very different monitoring needs.

In practice, the requirement for an EMP or monitoring plan to be submitted with the EIA is fairly common in the EA systems of low and middle income countries, and also in those of funding agencies (see the tables in Chapter 3). However, the requirement for an EMS to be submitted as well is much less common, even in high income countries, where it is a fairly recent innovation.

Even if an EMP and an EMS are called for, they may still be no more than good intentions, any more than the EIA, unless an independent check is made that the developer or operator is acting as intended. For the integrated framework shown in Figure 11.1 to be effective, the competent authority for EA must arrange for a supervisory level of monitoring or auditing to be carried out, over and above that done by the developer. Unless the competent authority for EA is the sole environmental authority, this will normally require close co-ordination with all other agencies with relevant responsibilities, such as the pollution control inspectorate, buildings inspectorate, waste management authority, water management authority etc. This need for close co-ordination over monitoring reinforces the need for effective consultation with other authorities in the earlier stages of the EA process, as discussed in Chapter 9.

Some developing countries and countries in transition have quite complex arrangements for environmental planning and management, with numerous authorities involved. This can make the task of co-ordinating a supervisory monitoring programme a complex one. In other less developed countries there may be just one newly-formed environmental agency responsible for all environmental matters. The problem is then not one of co-ordination, but of having the capacity to monitor everything that needs to be monitored. The environmental authorities in low and middle income countries are very often short of staff. A systematic approach, based on a framework similar to that outlined in Figure 11.1, can minimize the amount of monitoring that needs to be done by competent authorities, by clearly defining what is expected of the developer or operator.

11.3   Environmental Management Plans and Monitoring Plans

Although several developing countries and countries in transition include requirements for monitoring in their EIA procedures (Chapter 3), few are as explicit as they might be. Some development banks and aid agencies give rather fuller guidance however. In particular, World Bank policies (World Bank 1999) give a reasonably comprehensive definition of what should be included in an Environmental Management Plan. While these are specific requirements for World Bank funded projects, they are equally relevant to any large project. For smaller projects, a simplified interpretation of the same principles may be more appropriate. The World Bank's policy specifies that an EMP should include:

- *A summary of all potentially significant adverse impacts that are anticipated.* This can be taken directly from the EIA report
- *Full details of each planned mitigation measure, referenced to the anticipated impacts, including any environmental impacts of the mitigation measures, and linkages to any other relevant plans.* This too should be a summary of what is stated in the EIA report plus, where appropriate, design details, equipment descriptions and operating procedures. The measures covered should include all those which, if not undertaken, would lead to impacts greater than those predicted, whether or not the EIA report specifically describes them as mitigation measures
- *Monitoring and reporting procedures.* These should cover both the monitoring of impacts (to detect when particular mitigation measures may be needed) and the monitoring of progress of mitigation and its results. Reporting procedures should define in detail what monitoring results will be reported to the competent authorities, when, and in what form
- *Capacity development, training, and responsibilities for mitigation and monitoring.* While the developer has overall responsibility for all mitigation and monitoring, these measures cannot be carried out unless the developer specifies who will carry them out, and provides suitably trained staff. Responsibility may lie with the developer's own organization, or it may be the main contractor, or a particular sub-contractor. Often it may be necessary to specify a particular job function (e.g. the site manager) who is responsible for certain activities. Responsibilities should also be defined for any necessary training, supervision, monitoring of implementation, remedial action, financing and reporting
- *Implementation schedule and cost estimates.* This schedule should show how each mitigation measure is phased and co-ordinated with the overall project implementation plan. It is essential to demonstrate that all mitigation measures proposed have been fully costed by the developer before project approval, and that these costs have been included in the developer's financial appraisal of project viability
- *Integration of EMP with the project.* Responsibilities for implementing and supervising the EMP should be assigned within the overall project plan

Under World Bank policy, the capacity development component of the plan should also consider the needs of the competent authorities, as well as the developer, often under the Bank's own technical assistance programmes. This can apply to any agency funded project.

Many EA systems in low and middle income countries either make no reference to an EMP, or are fairly vague in their requirements. The more detailed requirements outlined above, or at least the principles which they reflect, may be regarded as good practice under any system. Further guidance on good monitoring practice has been provided by UNEP (UNEP 1996), which has prepared checklists for the structure and content of the monitoring plan which should form part of the EMP (Box 11.1).

The last two points in Box 11.1 emphasize the value of monitoring by other actors in the EA process than just the developer and the competent authorities. In

---

Box 11.1   Good Practice in EA Monitoring

Before approval establish responsibility for:

- Undertaking and paying for monitoring
- Managing the monitoring information
- Implementing any action required

Monitoring provides information on impact:

- Nature
- Magnitude
- Geographical extent
- Timescale
- Probability of occurrence
- Significance
- Confidence in prediction

Effective monitoring programmes:

- Have realistic sampling programmes
- Use relevant sampling methods
- Collect quality data
- Have compatibility of old and new data
- Have cost-effective data collection
- Are innovative
- Use appropriate databases
- Use multi-disciplinary interpretation
- Report internally and have external checks
- Respond to third party input
- Present data to the public

Source: UNEP (1996).

---

particular, both NGOs and the public at large have potentially a very valuable role to play in low and middle income countries, by extending the amount of time and effort devoted to monitoring very considerably, at no cost to the developer or the authorities (other than in taking remedial action). A bad EMP makes no reference to this potential source of information, while a slightly better one may describe a complaints procedure which investigates the validity of complaints received. A good one will show how the developer positively encourages information from these extra sources, for example by establishing a more responsive procedure for complaints and other information received, or by setting up local liaison committees and/or specific NGO monitoring programmes. Examples of involving NGOs and local people in monitoring programmes are given in Eckman (1996) and Rasid (1996).

In reviewing the environmental management plan, the competent authority should also pay particular attention to the proposed reporting procedures. It is through these that the authority becomes aware of whether or not mitigation measures have been carried out, of the results of monitoring, and of the actual impacts of the project, as identified by the developer, in accordance with the pre-defined plan. By ensuring, in consultation with other relevant authorities, that details are provided of how this information will be reported, in a thorough and timely fashion, the competent authority can minimize the amount of supervisory monitoring that it will need to do itself. An example of an EMP incorporating several of the above principles is summarized in Box 11.2.

Box 11.2   Greater Cairo Wastewater Project – Environmental Management Planning

The drilling of a main sewerage tunnel and feeder tunnels through a limestone rock outcrop north of Maadi in Greater Cairo entailed several potentially significant impacts (see the scoping matrix of Figure 4.2 in Chapter 4). Specific environmental management activities were defined within individual sections of the EIA report's description of proposed mitigation measures, and summarized in a mitigation management plan. A separate environmental monitoring plan was also drawn up. Responsibility for implementing these measures during construction was either incorporated into contract documents for sub-contractors or assigned to the main contractor's resident Engineer. Prime responsibility for monitoring of operational impacts was assigned to the wastewater authority, and it was recommended that health monitoring be conducted by the Ministry of Health. It was recommended that complaints be monitored and investigated systematically, although no procedures for doing so were defined. Provisions for control of noise and vibration, dust control practices and reinstatement of streets and footpaths were set out and included in contract documents, with requirements for monitoring and reporting of the results to the resident Engineer. Provisions were also set out for records of monitoring data and regular reviews to be submitted to the competent authority. To deal with possible archaeological impacts, procedures were defined for having an experienced archaeologist on call, and for continuous monitoring of particular activities. It was recommended that budget provision be made in the project costing for those monitoring and environmental management activities not covered by sub-contract provisions, although these costs were not detailed in the EIA report.

Reference: Taylor Binnie and Moharram-Bakhoum ACE (1997) *Greater Cairo Wastewater Project, Maadi Rock Tunnel Environmental Impact Statement* Taylor Binnie & Partners, Redhill, UK.

11.4   Environmental Management Systems

Many of the most serious environmental impacts that have resulted from development projects have arisen not because the project was badly designed, nor because it was badly implemented, but because it was badly managed in operation. The Bhopal and Chernobyl incidents are among the most catastrophic of these (to which bad design as well as bad operational management may have contributed), but there are many other examples. The impacts during construction and after commissioning may have been no greater than predicted, but a year later or 10 years later something has gone badly wrong, because the operation was badly managed.

The purpose of an environmental management system is to avoid this. At its simplest, an EMS consists of no more than a written description of how the operator's normal management procedures prevent significant adverse environmental impacts. Its benefit is that, in writing it down, the developer or operator is forced to think about whether or not his normal management procedures really do achieve that aim, and amend them if necessary.

At that simple level however, the EMS may be no more than good intentions, if the procedures it defines are not followed. To be effective, the EMS needs to be monitored, in the first place by the operator himself. This can be done through regular audits (known as internal audits), of the implementation of the relevant procedures, and of their effectiveness in avoiding significant adverse impacts. These are carried out by staff members independent of the activities being audited. However, to demonstrate to the world outside that the operator's management system really does achieve its aim, some independent external check is needed as well.

In part at least, this independent external check can and should come from inspection visits or audits carried out by the relevant environmental authorities. These visits may be made under pollution control legislation, waste management legislation, or other legislation, including environmental assessment legislation. Although such audits by the environmental authorities may be aimed primarily at ensuring that pollution and other consents are complied with, rather than at checking the adequacy of the operator's management systems, that may be all that is needed. If pollution control and other environmental legislation is comprehensive and is rigidly enforced, with heavy penalties for non-compliance, it is in the operator's own interest to manage the operation effectively, so as to avoid such penalties.

ISO 14001

It is largely for this reason that environmental management systems have been developed and applied extensively in the chemical and other potentially polluting industries in high income countries, and increasingly also in developing countries and countries in transition. In support of this, the International Standards Organization has issued the international standard ISO 14001, to provide an agreed definition of a sound EMS (ISO 1996).

ISO 14001 is one of a series of environmental standards issued by the ISO, covering matters such as life cycle assessment (ISO 14040) as well as the environmental management of operations. Among these, ISO 14004 (general guidelines on EMS), ISO 14010 (principles of auditing), ISO 14011 (audit procedures for EMS) and ISO 14012 (auditor qualifications) give additional support to the basic standard for environmental management systems, ISO 14001.

As well as defining what a sound EMS should comprise, ISO 14001 makes it possible for operators to obtain independent certification that their environmental management systems do indeed meet the requirements of the standard. The independent external check (in this case a management system audit) is then carried out by a nationally or internationally accredited certification body, at the operator's own expense. Many organizations throughout the world are increasingly calling for their suppliers of goods or services to have certification under ISO 14001, whether they be located in high income countries or low or middle income ones. To circumvent the pitfalls of weak accreditation procedures, they often make their own decisions on which certification bodies can be relied on to provide meaningful certificates.

The basic elements of an EMS under ISO 14001 are:

- A list of potential environmental impacts
- A set of operational procedures for monitoring, controlling and reducing the impacts, and recording the results
- A procedure for internal audits of the procedures

The first element can come directly from the EIA report. The second should be part of the operator's normal management procedures. The third is an extra management procedure needed to demonstrate that the rest of the EMS is working effectively.

An important feature of ISO 14001 is its requirement that impacts should not only be controlled, but reduced, with specific targets and action plans defined by the operator. This supports pollution control regimes which demand continual improvement from industrial operators, to use or move towards the use of best available techniques (BAT). This can be particularly important in transitional countries, where many industrial plants are old and highly polluting, but where immediate enforcement of high pollution standards could have severe economic implications. It can also be important in developing countries, as discussed below.

Integration of EMS with EIA and EMP

For major industrial projects, it may be appropriate to implement the framework outlined in Figure 11.1 rigorously, by requiring an EMS certified under ISO 14001 to be specified in the EMP, for approval with the EIA. This requirement has recently been included in guidance for the UK's oil and gas sector, and may well be equally applicable for similar projects in low and middle income countries. For other projects, obtaining the developer's commitment, in the EMP or EIA, to a less comprehensive EMS, can form the basis for the checks carried out during inspection visits or audits by the appropriate environmental authorities. An intermediate approach has been followed in Chile (Chapter 14), by requiring an independent environmental audit to be conducted at the developer's expense, and the results submitted to the authorities. At the very least, a clear specification of the monitoring that will be done by the operator, and of the records that will be kept, minimizes the amount of monitoring that needs to be done by over-stretched authorities.

Decommissioning

Depending on the complexity of the decommissioning process, it may be necessary for the developer to define, in outline at least, how the process will be managed, at the project approval stage. This will then form part of the operation's management procedures, although further refinement is likely to be necessary at a later date.

It may also be necessary, at the project approval stage, for the competent authority to obtain bonds or guarantees from the developer that funds will be available for decommissioning, which allow for the possibility of bankruptcy. Mining operations are typical of projects which are likely to fall in this category, in respect of site restoration and/or long-term protection of water courses.

11.5   Supervisory Monitoring and Enforcement

Implementation Monitoring and Ex Post Evaluation

Very few countries' EA systems include comprehensive provision for supervisory monitoring by competent authorities, to ensure that an EMP or its equivalent is actually implemented, and that impacts are no greater than predicted. The Taiwanese system includes some such provisions (Arts 1998), but even in the Netherlands, whose system is one of the most ambitious, ex post evaluation (an audit of the activity after commissioning) has been conducted for only a small proportion of EIAs. However, this does not mean that monitoring does not take place. In countries which have separate legislation for building regulation, pollution control and other related activities, much of the supervisory monitoring for EA that needs to be done by competent authorities may be carried out under the requirements of that other legislation. If monitoring reports are being provided by the developer in accordance with a sound EMP, and if there is close co-ordination between the competent authorities, a small number of spot-check visits by the appropriate authorities should be sufficient. These should be planned to take place at critical points in the construction/implementation programme, when certain key actions and construction impacts can be inspected, and when any necessary remedial actions can most readily be taken. A final post-implementation visit may also be necessary, to cross check the developer's report on monitoring of operational impacts.

In many low and middle income countries, the competent authority for EIA may be a newly created agency, with a shortage of trained staff, and ill defined overlaps between its responsibilities and those of other agencies. In these circumstances, it is essential to focus resources on those impacts which are of prime importance, and for which monitoring can make a real difference. Otherwise, monitoring can become an expensive academic exercise with no clear practical purpose (Arts 1998).

Most development banks and aid agencies have their own procedures for supervisory visits by their own staff. The World Bank, for example, undertakes general project supervision visits, and ex post project evaluation visits and reports, both of which should include examination of environmental aspects (World Bank 1996). If the agency's aims include capacity building of local institutions (which is normally the case), it is important that these visits be closely co-ordinated with the relevant local competent authorities, and with related technical assistance programmes.

The World Bank has identified environmental monitoring of the projects which it funds as an important weakness in its system, which it has taken steps to rectify (World Bank 1997, and Chapter 14). Of all category A projects (those requiring full

EIA) which were approved in fiscal years 1991–1994, only 12% had environmental specialists on all supervision missions, and 63% had no such participation at all. The performance of some other funding agencies may be even less satisfactory. An independent study of EIAs conducted in Tanzania revealed that only those conducted under local regulatory requirements (those of the national parks authority) were monitored during implementation or post-completion (Mwalyosi and Hughes 1998). The study found no examples of EIAs funded by donor agencies where the agency's interest had extended to ensuring that EIA recommendations were adhered to. As a result, in all but one of the donor funded projects examined, there was little evidence that EIA had made much difference to the actual environmental impacts of the project. As with monitoring by national agencies, there is a clear need for funding agencies to focus their efforts on those impacts whose monitoring can make a real difference to achieving the project's environmental and social objectives.

Sanctions

As noted in Box 11.1, it is essential to establish, before project approval, responsibilities for implementing any action that will be required as a result of monitoring. It is also essential to establish what sanctions will be applied if necessary, with appropriate legal backing. In the case of impacts that are subject to specific legislation, such as pollution control, the enforcement provisions of that legislation may suffice. Where no such legislation exists, or for impacts not covered by it, sanctions may need to be provided for in the EIA approval. This is normally done by attaching a set of conditions to the approval, as discussed in Chapter 10, which identifies all those mitigation measures and potential impacts that are considered to be particularly important. The relevant EA legislation should define the sanctions which may be applied if those conditions are breached. To minimize the possibility of legal disputes over liability, binding commitments may also be included in the relevant contract documents for the project, e.g. the contract between the developer (particularly if this is a public body) and the main contractor. As well as defining specific mitigation actions, these provisions may state what further action the contractor must take if an impact is found to exceed the prediction.

For developments in low and middle income countries, the developer or project sponsor is often a public authority. This can cause difficulties in applying sanctions, particularly when the political will to enforce EA is weak. When a development bank or aid agency is involved it can overcome this problem by applying its own sanctions. Environmental conditions resulting from the EA need to be written into the funding agreement and appropriate contracts, and the funding can be phased, so that payments can be withheld until the conditions are met.

However, although sanctions are an essential backup, the most important requirements for effective enforcement of EA are clarity in what is expected of developers and contractors, close communication with them, and checks to ensure that they do indeed provide what is expected. The EIA report by itself may not be sufficient for this, and often needs to be supported by clear approval conditions, contract conditions, an agreed EMP, and supervisory monitoring of the EMP.

11.6 Environmental Audit

Regulatory Inspections and Audits

For projects such as roads or commercial or residential buildings, the main operational impacts occur as a direct result of project construction and implementation, and little if any ongoing monitoring is necessary, beyond an ex post evaluation. For others, including potentially polluting industries, some of the impacts can change significantly if the operation is not properly managed. For these types of project, and these types of impact, the operators' management systems (or EMS), and/or the actual impacts, may need to be subjected to some form of ongoing independent inspection or audit. A fully certified EMS can provide this, at least in part, but in general at least some degree of inspection or auditing needs to be done by the relevant competent authorities.

Additionally, the growth of industrial activity common in many developing countries cannot take place without growing pollution, unless the pollution from existing industries is reduced. This is made feasible by advances in technology, and regulatory mechanisms which require operators to update their facilities to use the best available techniques (BAT). As well as carrying out periodic inspections or audits to ensure that an installation is operating as approved, the competent authority needs to periodically renegotiate the approval itself, or the pollution permit. Furthermore, in order to reduce pollution while still allowing new industries to be built, the auditing and permitting processes have to be applied to existing industries as well as to new ones.

High income countries generally have well established pollution control systems of this nature, which are often partially or entirely separate from EIA. In many low and middle income countries, both EIA and pollution controls are new, and the authorities have the task of applying environmental controls to previously unregulated old industries, as well as to new ones. In some countries, particularly those transitional economies which are aiming for harmonization with European Union legislation, it may be appropriate to adopt a similar approach to those developed in high income industrial countries, such as the EU's integrated pollution prevention and control directive (CEC 1996). However, high income countries' systems tend to be fairly complex, and simpler approaches may initially be more appropriate in low and middle income countries. In some countries such as India, specific legislation has been introduced for environmental audit, which is applied to the pollution control of existing industrial activities. This is linked to but separate from EIA legislation, which is used as a planning control for a wider variety of activities. In other countries such as Bulgaria, EIA legislation applies to both new and existing activities, to cover both planning approval and pollution control. In both cases, a distinction needs to be drawn between the broad scope of EIA as a planning control, applied mainly to medium or large scale projects, and the narrower scope of environmental audit, primarily as a pollution control, applicable in principle to any size of project.

For medium and large scale industries, regulatory environmental audits rely to a large extent on checking the management systems and records maintained by the operator (whether or not these can be classed as an EMS), together with physical inspections. For small ones they rely more heavily on inspection, plus whatever direct environmental measurements are feasible within time and budgetary constraints. Inspections and measurements are often carried out in more depth when a complaint has been received.

While pollution impacts are the main ones for which monitoring and auditing are likely to be required throughout the life of a project, others might be identified within individual EIAs. Seismic risks are one example, while health impacts may need to be monitored for several years after implementation (both are discussed in Chapter 5). In such cases, the EIA and/or EMP should clearly define the developer's responsibilities, to which authorities the developer will be responsible, and under what legislation or procedures.

Other Forms of Environmental Audit

In addition to regulatory audits carried out by competent authorities, the internal and external management system audits carried out under ISO 14001 are two of several types of environmental audit that may be carried out for various different purposes. Other types include waste minimization audits, compliance audits (for compliance with legal requirements) and due diligence or liability audits (for potential legal liabilities, particularly when the operation is being bought, sold, extended or privatized). All of these are generally carried out by operators or investors, or consultancies acting for them.

Under the operating procedures of development banks such as the World Bank (World Bank 1995) and the European Bank for Reconstruction and Development, liability audits are often associated with environmental assessment, when the proposed activity involves changes or extensions to existing facilities. The aim of the audit is to ensure that the bank is aware of any environmental liabilities from the existing operation (such as contaminated land) before making its decision on any new investment.

11.7   Monitoring in Strategic Environmental Assessment

One particular form of strategic environmental assessment that has been applied in developing countries, particularly when international funding agencies are involved, is programmatic environmental assessment (World Bank 1993, and Chapter 14). This applies when the agency's funding decision is made for an entire programme, on the basis of a single EA. At the approval stage, it is unlikely that sufficient information will be available to carry out the assessment in the level of detail that will ultimately be required. For some critical sub-projects, a detailed EIA may subsequently be called for, but the details for the rest of the programme would be established through the EMP produced with the programmatic EA. This approach is particularly common for major road schemes (Box 11.3). In effect, it replaces project level EIA by an EMP, and so the EMP needs to be extremely sound, to avoid reducing the effectiveness of the assessment.

Box 11.3   Environmental Management Planning in Programmatic EA

The route from Manyoni to Kigoma in Tanzania is nearly 1000 km long. At the stage of conducting the feasibility study for a proposed programme to improve the road, fully detailed plans were not available, and a fully detailed environmental assessment was not practicable. Instead, the EA was based on the outline proposals, and an environmental management plan was drawn up to define how impacts would be studied in detail during implementation, and mitigated as necessary. A fundamental part of the plan was the recruitment, training and management of an environmental management team to conduct these activities. Training for local decision-makers was also proposed, together with the formation of an advisory group, and a mechanism for evaluating individual impact studies. Environmental stipulations were defined for incorporation into sub-contract tender documents. Methods of involving local communities in the design of mitigation measures were devised, and also plans for co-ordination between the various authorities involved. A monitoring plan was drawn up, which included monitoring contractors' responsibilities for borrow-pit rehabilitation, landscape revegetation, erosion and sedimentation control, waste management and cultural protection activities. The costs of these activities were included in the construction budget, and the costs of the environmental management team were detailed separately for inclusion in the overall financial appraisal of the project.

Reference: Gannett Fleming and Co-Architecture (1995) *Manyoni-Kigoma Road Feasibility Study Draft Final Environmental Report* Gannett Fleming Inc (USA) and Co-Architecture (Tanzania), Dar es Salaam, Tanzania.

The EMP produced for such a programme is basically as described in Section 11.3, but certain aspects assume particular importance. The mitigation measures for the programme must include detailed assessment of the environmental baseline for each sub-project that is not subject to a full EIA, and of the potential impacts of alternative options (e.g. alternative routings, alternative sources of materials etc.). Mitigation responsibilities defined in the EMP must include the responsibilities for carrying out these mini-environmental assessments, for training staff who may undertake them, for involving the public, for reviewing the results, for making decisions, and for monitoring the subsequent activities and impacts. It is also essential that all these activities be fully costed in the EMP, and included in the overall programme budget.

At the level of policies and plans, SEA experience is more limited. A number of ad hoc SEAs have been carried out, while others have been conducted under the screening procedures of national EA systems (e.g. in Slovakia) or those of international agencies (e.g. the World Bank). However, no general monitoring practice has yet emerged. This applies to National Environmental Action Plans, Sustainable Development Strategies and other specific environmental policy instruments, as well as to the environmental assessment of other policies and plans. However, a similar framework to that in Figure 11.1 may be applied, as shown in Figure 11.2.

For most policies and plans, there is no operational phase distinct from implementation. Implementation begins when the policy or plan is agreed and put into effect, and continues until the policy or plan is next revised. The effects are unlikely to be instantaneous, but will evolve throughout implementation, and so implementation monitoring needs to take place for the life of the policy or plan.

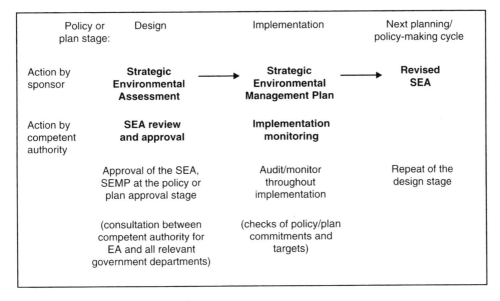

Figure 11.2   *SEA Monitoring and Management*

As with projects, the prime responsibility for monitoring should rest with the sponsor of the policy or plan, backed up by checks by the competent authority for EA, or some other independent body. For internationally funded SEAs, the funding agency should carry out its own checks in parallel with those of the local competent authority, although they are unlikely to be of the same duration. The sponsor's monitoring programme, together with a detailed implementation programme for the environmental aspects of the policy or plan, should be defined at the approval stage, in a management plan equivalent to a project's EMP. The content of this strategic EMP should be similar in principle to that described for projects in Section 11.3.

In the case of National Environmental Action Plans and other environmental policy instruments, the sponsor is often the country's environment ministry. The ministry's ability to monitor implementation independently will be limited, unless the plan is successfully embedded in other ministries' development policies and plans.

11.8   Conclusions

Some form of follow-up monitoring after the approval stage of EA, together with measures for taking corrective action when needed, are essential for the effectiveness of the process. Achieving that apparently straightforward aim can be extremely complex.

National environmental agencies are often short of staff, with ill defined divisions of responsibility between themselves and other relevant government departments. For funding agencies the issues are somewhat less complex, but they too have had great difficulty in implementing effective follow-up monitoring and enforcement.

Part of the solution to these problems can come from placing clear responsibilities on developers and operators, through a structured approach to the use of environmental management plans and environmental management systems, linked directly to EA. However, it is still necessary for the agencies to monitor the implementation of any such EMP or EMS. In order to do so effectively, resources need to be focused on those impacts of highest priority, and for which monitoring can make a significant difference. Follow-up programmes need to be adapted to the specific priorities of individual countries, and the specific priorities of individual projects.

The priorities for funding agencies and national authorities are not necessarily the same. For example, a funding agency's goals may be closely related to the social impacts of development, while the responsibilities assigned to a national environmental agency may relate more to pollution control. It has been clearly recognized by international agencies that technical assistance in the form of institutional strengthening and capacity building can help to address both sets of issues. However, it is vital that the two sets are not confused. The institutional structures and capacities needed for follow-up of social impacts, health impacts, and other impacts identified in an EIA, may be very different from those needed for effective pollution control. In designing technical assistance programmes, careful consideration has to be given to their aims, and to the structures most suited to achieving them.

Finally, it must be recognized that even when responsibilities are placed on developers and operators, the cost of establishing and maintaining institutions capable of effective monitoring and enforcement is high. International agencies can provide initial assistance, but for development to be sustainable, the bulk of the cost must be borne by national governments. To raise the necessary revenues the polluter pays principle may be applied, for example through pollution levies and permitting charges, linked directly to the regulatory agencies' costs. To demonstrate the need for such charges, the relative costs of pollution and its prevention need to be analysed and understood at the highest levels of government. Funding agencies can contribute to this too, by including the development of appropriate environmental economics expertise in their technical assistance programmes.

Economists, legislators, investors, developers, operators, EA practitioners, NGOs and the public as a whole, as well as government authorities, international banks and aid agencies, all have important roles to play. By helping to improve monitoring, enforcement and impact management, they can make a major contribution to improving the effectiveness of the EA process as a whole

Discussion Questions

1. Why is monitoring important? What are the main barriers to effective monitoring?
2. How can environmental management plans contribute to effective implementation of EIA? What are their limitations?
3. What actions can be taken by funding agencies to help improve monitoring and enforcement?

4. For what types of project is a fully certified EMS under ISO 14000 most appropriate? How can competent authorities encourage good environmental management throughout the life of other projects?

Further Reading

Guidance on monitoring, implementation and auditing within the EIA process in developing countries and countries in transition is given in UNEP (1996). Arts (1998) gives an overview of ex post evaluation in national EIA systems, and a comprehensive review of the system in the Netherlands. Environmental auditing, environmental management systems and other aspects of environmental management are described in Welford (1996). An overview of World Bank experience is given in World Bank (1997), and suggestions for future developments in Goodland and Mercier (1999).

References

Arts, J (1998) *EIA Follow-up: On the Role of Ex Post Evaluation in Environmental Impact Assessment*, Geo Press, Groningen

Commission of the European Communities (CEC) (1996) Council Directive 96/61/EC concerning integrated pollution prevention and control, *Official Journal of the European Communities* L257/28

Eckman, K (1996) How NGOs monitor projects for impacts: results of recent research, *Impact Assessment* **14**: 241–268

Goodland R and Mercier JR (1999) The Evolution of Environmental Assessment in the World Bank: from 'Approval' to Results *Environment Department Paper No 67*, World Bank, Washington DC

ISO (1996) *Environmental Management Systems – Specification with Guidance for Use*, ISO 14001:1996(E) International Standards Organization

Mwalyosi, R and Hughes, R (1998) *The Performance of EIA in Tanzania: an Assessment*, Institute of Resource Assessment, Dar es Salaam, and International Institute for Environment and Development, London

Rasid, H (1996) Impact assessments from survey of floodplain residents: the case of the Dhaka-Narayangan-Demra (DND) project, Bangladesh, *Impact Assessment* **14**: 115–132

UNEP (1996) *Environmental Impact Assessment Training Resource Manual*, United Nations Environment Programme, Nairobi

Welford, R (1996) *Corporate Environmental Management*, Earthscan, London

World Bank (1993) *Sectoral Environmental Assessment Environmental Assessment Sourcebook Update No 4*, World Bank, Washington DC

World Bank (1995) *Environmental Assessment Sourcebook Update No 11: Environmental Auditing*, World Bank, Washington DC

World Bank (1996) *Environmental Assessment Sourcebook Update No 14: Environmental Performance Monitoring and Supervision*, World Bank, Washington DC

World Bank (1997) *The Impact of Environmental Assessment: A Review of World Bank Experience*, World Bank, Washington DC

World Bank (1999) *Operational Policy OP 4.01 Annex C: Environmental Management Plan*, World Bank, Washington DC

PART TWO

Country and Institutional Studies of EA Procedures and Practice

12

Country Studies of EA in Chile, Indonesia and the Russian Federation

12.1 EIA IN CHILE

Luis C. Contreras

1 Introduction

The EIA system in Chile is embodied within the Environment Framework Law (EFL), No. 19.300, which was enacted in March, 1994. The EFL establishes a transectoral system of environmental management which crosses every governmental agency and sectorial activity; and recognizes the existence of numerous environmental competencies spread throughout the government administration (Asenjo 1994; Aylwin 1994; Contreras 1996). Consequently, the function of the government environmental institution established by the EFL, the National Commission for the Environment (CONAMA), is of a coordinating nature. At the national level it is formed by: (i) a decision-making Directive Board compromising 13 ministries, chaired by the Ministry of the Presidency; (ii) a technical advisor and coordinating Executive Director; and (iii) a Consulting Committee with two representatives from the scientific community, environmental NGOs, academic institutions, private productive sector, workers' organizations, and a representative of the President of the Republic. A similar arrangement exists in each administrative region, where the Regional Commission for the Environment (COREMA) is chaired by the Regional authority (the Intendente) with the addition of a Technical Committee formed by the Regional Directors of the Regional Governmental Agencies with environmental competencies.

Environmental Assessment in Developing and Transitional Countries. Edited by N. Lee and C. George.
© 2000 John Wiley & Sons, Ltd.

The EIA system created by the EFL became mandatory in April, 1997 with the enactment of an EIA regulation (MSGP 1997). Before that, from October 1993, a voluntary system of EIA operated based on a Presidential Decree (CONAMA 1993; Espinoza *et al.* 1994a). This established procedures on how to submit and review the growing number of Environmental Impact Studies (EIS) that developers were voluntarily submitting to the authorities. It only considered the submission of EISs and excluded the possibility of submitting affidavit Environmental Impacts Declarations (EID); a second type of environmental assessment document for projects whose impacts are not considered significant. It also excluded public participation procedures. On the other hand, it did include the establishment of Terms of Reference (ToR), which is not formally included in the EFL.

2   Main Features of the Chilean EIA system

Screening

The EFL lists a great number of types of projects and activities subject to the EIA procedure, making no distinction between private and Government led projects. Military installations are subject to their own standards within the objectives of the EFL. The 1997 Regulation characterizes to a greater extent than the EFL the projects and activities to be subject to EIA procedures. However, this characterization does not have the same level of quantitative specification for projects in different economic sectors (Ormazábal 1997a). Consequently, some of these projects, even the smallest ones, are formally subjected to EIA. For cases where a quantitative threshold has been established, projects or activities below this are not subject to EIA, based on the assumption that they do not have significant impacts. A first modification to the Regulation was introduced in September 1998 by specifying the minimum size of urban projects subject to EIA procedures.

Projects should be submitted for evaluation only once and should provide sufficiently detailed information to assess the potential impacts. Developers are not required to submit alternatives of location or technology to be used. Normally, submission occurs no earlier than conceptual or feasibility engineering studies, a time at which is difficult to introduce strategic modifications (Ormazábal 1997a).

Scoping

If a listed project complies with environmental regulations and is not expected to result in significant impacts according to a set of environmental criteria, a developer may submit an EID indicating that such is the case. The criteria used refer to effects on: (i) human health; (ii) natural resources; (iii) protected areas; (iv) traditional cultures; (v) landscape; (vi) touristic values; and (vii) cultural heritage. If a project complies with the existing regulations, but not with all the indicated criteria, a developer must submit an EIS.

The present mandatory EIA system does not require the negotiation of ToR for EISs, which were specified under the Presidential Decree during the voluntary phase. The absence of a clear definition of the minimum, sufficient and necessary information required to assess impacts in any particular case is, in the author's view, the most serious impediment to an effective and efficient functioning of an EIA system (Contreras 1996, 1998; Ormazábal 1997b). Often the authorities ask for relatively large amounts of irrelevant information for the base line studies. Since such requests for base line information occur during the (relatively late) review stage, developers and environmental consultants preparing EIS usually include a wide range and amount of information in order to prevent the project from being delayed by requests for information which otherwise may not be available. Consequently, EIS's can become unnecessarily long and require large amounts of time, human and financial resources to prepare and review them. This is a common and enduring problem in the international practice of EIA (Wood 1995). Strictly, according to the EFL and the EIA Regulation, EISs should only address the key issues that justify the need to submit an EIS.

The lack of a comprehensive set of environmental quality standards in several cases, as well as the lack of explicitly defined environmental policies, creates an increasing number of conflicts regarding the levels of significance and/or acceptance of certain impacts, e.g. relating to native forest exploitation. Environmental quality standards are being developed following a procedure established by the EFL and a supporting Regulation (MSGP 1995).

EIS Content

EISs should include:

1. A description of the project, including the post closure (decomissioning) phase.
2. The pertinent environmental legislation, indicating the project's compliance with its requirements.
3. The reasons why submission of an EIS is required according to the regulation.
4. A base line study (focusing on the environmental components related to the criteria which gave rise to the requirement to submit an EIS), covering physical, biological, socioeconomic, infrastructure, land use, historic heritage, landscape and risk conditions.
5. An assessment of impacts; most commonly done using a modified method based on the Leopold Matrix.
6. A mitigation, recovery and compensation plan. Compensation should be in the same 'environmental currency' as the impacts. However, in some cases, negotiations parallel to the EIA process between the developer and the local community have included non-environmental compensations, such as investments and programmes favouring social needs.
7. A monitoring plan of relevant variables.
8. An account of public participation activities undertaken before submission of the EIS. The EIA system only requires public participation once the EIS has been submitted but sometimes this is implemented voluntarily by the developer at an earlier stage.

9. An appendix should contain more detailed supporting information, such as laboratory reports, a list of the professional authors of the EIS etc.
10. An executive summary.

EID and EIS are usually prepared by consultants hired by the developer.

The Authorities and Technical Reviews

An EID or EIS must be submitted for review and decision to COREMA if the environmental effects of the project are confined to one administrative region, or to CONAMA if they affect two or more regions. From north to south, Chile has 13 administrative regions.

EIDs and EISs are reviewed by all Government Agencies with competence over the affected environmental components. There is a 60 working day period, including Saturdays, extendible by 30 days, to review an EID; and 120 and 60 days, respectively, to review an EIS. If, at the end of this period, there is no response by the Authority, the permit is considered as granted in the case of EISs. If the authority requests the developer to provide further explanations or complementary information, the formal review period is stopped and is resumed when the request is satisfied.

Technical reviewing by the governmental agencies consists of verifying that the EIS satisfactorily covers the Agencies environmental concerns. An integrated review is undertaken by the CONAMA Regional Director Office, which may hire external consultants for their assistance. A review methodology guideline based on the hierarchical method of Lee and Colley (1992) published by CONAMA (CONAMA 1993; Espinoza *et al.* 1994a) has been used infrequently.

Environmental Insurance Policy

A novel feature of the Chilean EIA system is the possibility to submit an EIS together with an insurance policy covering the risk of damage to the environment if construction is initiated during the reviewing period. In such a case, a provisional authorization may be obtained to initiate the project, without being bound to the final decision. Since the enactment of the EIA Regulation – up to October 1998 – 12 EISs have been submitted with an insurance policy, and three have been accepted. There is no specified period of time in which the authorities must accept or reject the insurance policy.

Public Consultation

During a period of 60 days after the publication of an abstract of the EIS in newspapers, legal representatives of citizens' organizations (NGO and others) and individuals directly affected by the project may acquaint themselves with the EIS and formally submit their concerns. Public concerns formally submitted must be taken into account by the authority in its final resolution. Besides this basic mechanism established by the EFL, it is the responsibility of the COREMA to promote public participation. This usually includes an oral presentation of the EIS by the developer to the local community.

Authorization

The Regional Director of CONAMA, as Executive Secretary of the COREMA, compiles and synthesizes the different government agencies reports and, after also considering the opinion of the public, prepares a Technical Review Report to inform the COREMA. The COREMA makes a final decision on granting or denying the environmental authorization for the project. An important conceptual point is that EISs are prepared to make sure that projects or activities comply with the existing legislation and, mainly, to assess the impacts on issues that do not or cannot have standards, e.g. relocation of population. Consequently, granting or denying authorization is a political decision of the COREMA relating to non-regulated environmental impacts considering, among other things, the Technical Review Report.

In the Chilean system the authorities do not have the power to change projects, but they do have the power to set conditions that must be satisfied by the developer. This partially explains why the Environmental Qualification Resolution (RCA) for some projects may become quite long (over 150 pages in some instances). Approving conditions mainly focus on mitigation, contingency and compensation, and monitoring plans.

Upon approval of the EIS all other technical permits of an environmental nature for a project are granted pending the submission of the required specific technical information. The Regulation lists a total of 30 environmentally related permits which concern activities in ocean coastal areas, aquiculture, national monuments, subterranean water, nuclear activities, mining activities, drinking water, waste water, solid wastes, environmental risk and soil conservation.

Monitoring

The monitoring plan usually includes provision for at least an annual monitoring report to the authorities. Nevertheless, it is the responsibility of various government agencies to enforce compliance with the law and the conditions stated in the RCA. So far there has not been a general evaluation of compliance with approval conditions.

Interestingly, a growing number of projects are being approved with the condition to make provision for independent environmental audit programmes. Auditors are chosen by COREMA/CONAMA from three alternatives proposed by the developer. Auditors must report to both the developers and COREMA/CONAMA. Auditing costs are covered by the developer.

Appeals

In every instance of the EIA process, any stakeholder may appeal to the superior administrative level, or to the Judiciary system, for a judgement over a dispute.

During the voluntary phase of the EIA system, seven cases out of 186 were submitted to the courts. Only one of these was resolved against the promoter of the project partly due to the weak legal support for the EIA system during its voluntary phase and the lack of an EIA Regulation until April 1997.

Conflicts

Besides the usual conflicts of interest and related actions by the local community or NGOs, probably the most serious situation occurs when the local community and some NGOs have physically impeded field measurements for base line studies. This situation may reflect poor early management of public participation by the developer.

3 EIA Practice in Chile

The Transitional Voluntary EIA System

During the three and a half years of the transition period a total of 186 projects, involving over US$23 405 million of investment, were submitted, mostly in the mining (41%), energy (22%), industry (17%) and sanitation (12%) sectors (Espinoza *et al.* 1997). Most projects have had an investment of over US$1 million (81%), with 26% of projects being in the range of US$100 million.

Projects were concentrated in the northern Regions II (25%) and III (13%), and the central VIII and Metropolitan Regions (12%, each). However, most investments, relating to huge mining projects, have occurred in three northern Regions (53%). Of the projects, 7% have involved more than one administrative region in central Chile.

Up to April 1997, out of 186 projects submitted during the voluntary phase, 102 had been approved by the authorities, five had been rejected (all of them related to sanitary landfills in the Santiago Metropolitan Region) and the remainder were under revision. Most projects submitted during the voluntary phase originated from the private sector. Since the enactment of the Regulation the proportion of public sector projects has been increasing.

The relative success of the voluntary system, derives partly from the fact that it allowed developers to obtain an official approval required by financial institutions or investors. Also, it enabled developers to obtain environmental approval for several sectoral permits by dealing with only one co-ordinating authority, rather than with several of them acting in an independent and sometimes contradictory manner.

The Mandatory Phase

The enactment of the Regulation has involved several important changes to the EIA system. These include: (i) establishment of a mandatory system for public and private projects; (ii) the possibility to submit EIDs; (iii) the possibility to start projects before obtaining the RCA if an environmental insurance policy is accepted; and (iv) the mandatory submission of urban zoning plans to assessment.

In the first 20 months of the mandatory phase 86 EISs and 696 EIDs have been submitted. Most private sector projects requiring EIS were already being submitted in the voluntary phase and their number should not increase significantly. Indeed, their number may even decrease considering the alternative to submit an EID. Public sector projects submitted to the EIA system should increase in comparison with the voluntary phase.

After 20 months of application of the Regulation a proposal for modifications is being discussed. These are aimed mainly at clarifying and specifying the original version.

Strategic Environmental Assessment (SEA) is not yet undertaken in Chile. The only activities subject to EIA, which come close to SEA, are urban zoning plans (Glasson *et al.* 1999; Wood 1995). During the voluntary phase none of these urban plans was submitted. Indeed, the authority tried to postpone their submission in the belief that it was unnecessary to apply an environmental tool (EIA) in addition to an existing management tool (urban plans). It was also argued that there was insufficient experience in environmentally assessing urban zoning plans, and/or that there were legal and practical incompatibilities in the application of EIA procedures to them. This attitude derives mainly from the Ministry of Housing and Urban Planning wishing to retain their decision-making power free from any external interference.

Support for the Implementation of the EIA System

The most important financial support for the implementation of the EIA system has come from a CONAMA/IRDB (InterAmerican Reconstruction and Development Bank) project for environmental institutional development. The project had a total budget of US$32.7 million, of which US$21.2 million were allocated by the Chilean Government and US$11.5 million by IRDB (CONAMA 1997a). The project started in 1993. Although all the 10 programmes within the project included some provisions relating to the implementation of the EIA system, there was one specific programme devoted to it with an allocation of US$7 million. It provided support to: (i) establish environmental technical units at CONAMA, COREMAs and ministries; (ii) promote EIA procedures in the public and private sectors; (iii) train public sector personnel; (iv) develop environmental guidelines; and (v) define environmental standards. CONAMA has published several documents related to EIA (Espinoza *et al.* 1994a,b; TESAM S.A.-CONAMA 1996; CONAMA 1997b).

4   Conclusions

The overall assessment of EIA practice in Chile is definitively positive. Some of the strengths of the Chilean EIA system include:

- Existence of one standardized EIA procedure for all the country and sectors
- Participation of all the government agencies in a process coordinated by one institution, CONAMA/COREMA

- Evaluation of and decision on, EIS and EID at the regional level, regardless of the national or strategic relevance of the project, contributing to the decentralization process
- Procedural alternatives for projects with and without significant environmental impacts, i.e. EID and EIS
- Public participation is ensured at least once an EIS is submitted to the authorities

Some weakness in the system, which may be improved in the future, include:

- Incomplete coverage of sectorial environmental policies
- Incomplete coverage of environmental standards
- Uneven quantitative criteria to determine projects subject to EIA
- Demand for irrelevant information by some governmental agencies
- EIS submission restricted to a late stage in project development making it difficult to introduce strategic modifications
- Unclear definition of roles and legal competencies for some government agencies
- Hesitance of some government agencies to be coordinated by CONAMA/COREMA
- Lack of public participation in the early stages of project development
- Administrative and coordination failure due to lack of human, infrastructure and financial resources allocated to environmental management in government agencies, including CONAMA/COREMA
- Deficient management of conflicts of interest by the developers and authorities in relation to controversial and problematic projects
- Need to implement SEA for higher levels of activities
- Need to develop guidelines for the mandatory and non-mandatory phase of the EIA procedure, e.g. environmental assessment and public participation in the early phases of project development
- Need for formal recognition of consultants participating in the EIA process

References

Asenjo, ZR (1994) Intervención del Secretario Ejecutivo de la Comisión Nacional del Medio Ambiente (CONAMA), Sr. Rafael Asenjo Zegers, in *Ley de Bases del Medio Ambiente*, CONAMA

Aylwin, AP (1994) Texto del Mensaje de SE el Presidente de la República, Don Patricio Aylwin Azocar, con el que Envió al Congreso Nacional el Proyecto de Ley sobre Bases Generales del Medio Ambiente, in *Ley de Bases del Medio Ambiente*, CONAMA

CONAMA (1993) *Instructivo Presidencial Pauta para la Evaluación del Impacto Ambiental de Proyectos de Inversión*, CONAMA

CONAMA (1997a) *Gestión Ambiental del Gobierno de Chile*, CONAMA

CONAMA (1997b) *Orientaciones para la Evaluación de Impacto Ambiental de Proyectos de Producción, Almacenamiento, Transporte, Disposición o Reutilización de Sustancias Tóxicas, Explosivas, Radioactivas, Inflamables, Corrosivas o Reactivas*, CONAMA

Contreras, LC (1996) Recent EIA developments in Chile, *EIA Newsletter* **13**:10

Contreras, LC (1998) Biodiversidad y el Sistema de Evaluación de Impacto Ambiental en Chile *Derecho del Medio Ambiente*, Editorial Jurídica Cono Sur Ltda Santiago, pp. 19–35

Espinoza, G, Pisani, P and Contreras, LC (eds) (1994) *Manual de Evaluación de Impacto Ambiental Conceptos y antecedentes básicos*, CONAMA

Espinoza, G, Pisani, P, Contreras, LC and Camus, P (eds) (1994a) *Perfil Ambiental de Chile*, CONAMA

Espinoza, G, Garcia, S, Valenzuela, F and Jure, J (1997) Algunas Experiencias Derivadas de la Aplicación del Sistema Voluntario de Evaluacion Ambiental en Chile, *Documento de Trabajo No 35*, Centro de Estudios del Desarrollo (CED)

Glasson, J, Therivel, R and Chadwick, A (1999) *Introduction to Environmental Impact Assessment*, 2nd edn, UCL Press, London

Lee, N and Colley, R (1992) Reviewing the quality of environmental impact statements, *Occasional Paper 24* 2nd edn, EIA Centre, University of Manchester, UK

Ministerio Secretaría General de la Presidencia (MSGP) (1995) *Decreto No 93 Reglamento para la Dictación de Normas de Calidad Ambiental y de Emisión*, Publicado en el Diario Oficial el 26 de Octubre de 1995

Ministerio Secretaría General de la Presidencia (MSGP) (1997) *Decreto No 30 Reglamento del Sistema de Evaluación de Impacto Ambiental*, Publicado en el Diario Oficial el 03 de abril de 1997

Ormazábal, C (1997a) Sistema de Evaluación de Impacto Ambiental (I): Análisis crítico del marco legal y su reglamento, *Ambiente y Desarrollo* **13**: 47

Ormazábal, C (1997b) Sistema de Evaluación de Impacto Ambiental (II): Análisis crítico desde la perspectiva de los actores y su responsabilidad en los EsIA, *Ambiente y Desarrollo* **13**: 51

TESAM S.A. – CONAMA (1996) *Metodologías para la Caracterización de la Calidad Ambiental*, CONAMA

Wood, C (1995) *Environmental Impact Assessment: A Comparative Review*, Longman, Essex, UK

12.2   EIA IN INDONESIA

Zulhasni

1   Introduction

The Indonesian EIA system, known as AMDAL, is an integrated process for coordinating the planning and review of proposed development activities. It complements the evaluation of the project's technical and economic feasibility. In Indonesia, the concept of EIA is seen as a way to achieve sustainable development because it may assist the project proponent in ensuring the development of his business in a manner which minimizes any negative environmental impacts and makes it socially acceptable. From the perspective of the decision maker authorizing the project, AMDAL is a tool for considering the impacts of a proposed activity on the environment and to act as early as possible in order to mitigate any negative effects which it may cause and to provide alternative solutions.

There are four types of AMDAL applied in Indonesia:

- EIA for an individual activity which has the potential to affect significantly the environment
- EIA for various integrated/multisectoral activities which is required for a project or business which involves more than one authorized government agency and where the activities are located in a single ecosystem type

- EIA for a special area which contains a number of development projects within it (for example, an industrial estate, tourism development area or a special bonded import and export area)
- EIA for a regional area which covers businesses or activities which are located in one zone of a regional development planning area and which has been defined in accordance with the general spatial plan for the region

Over the 12 years of its existence up to 1999, the implementation of AMDAL in Indonesia has had a number of positive effects, especially in increasing environmental awareness among bureaucrats and industrialists. However, some problems exist which have hampered its implementation, particularly a lack of understanding among participants that AMDAL should be used as a planning tool, the preparation of poor quality documents and the lack of monitoring of environmental management plans (RKLs) and surveillance of environmental monitoring plans (RPLs).

2   EA Regulations

The AMDAL process was established in 1986 by Government Regulation No. 29 which was promulgated under Law no. 4 of 1982, Indonesia's fundamental environmental law. This regulation applies to 14 central sectoral governmental departments and to 27 provincial governments. Each ministry or provincial government is required to develop sectoral AMDAL procedures and guidelines consistent with the basic guidelines issued by the Ministry of Environment and to establish an Environmental Review Commission. Overall co-ordination of the AMDAL process rests with BAPEDAL, the Environmental Impact Management Agency, which is a division of the Ministry of Environment.

According to Regulation No. 29, projects that were already under way or under construction in 1987 were subject to review, if their environmental impacts had not already been assessed. This process, known as SEMDAL, involved the preparation of an environmental evaluation study. However, new projects, that were expected to have a significant impact on the environment, were required to follow the AMDAL process. The significance of an impact is to be determined according to: (a) the number of people affected; (b) the extent of impact; (c) the duration of the impact; (d) the intensity of the impact; (e) the number of other environmental components affected; (f) the cumulative nature of the impact; and (g) the reversibility or irreversibility of the impact.

Both AMDAL and SEMDAL processes are designed to result in the preparation of environmental monitoring and management plans. In principle, these plans provide an operational basis for implementing and enforcing the findings and conditions that emerge from the AMDAL/SEMDAL review process. In practice, however, the benefits of these innovative features have not yet been fully realized because there have been too many existing projects to review and no standards upon which to base their revision nor any mechanism to monitor their compliance.

In October 1993, the government issued Regulation No. 51 to eliminate some of the confusion and delays which arose under the procedural guidelines formulated in

Regulation no. 29. The new regulation simplifies the initial screening process, eliminates the preliminary impact assessment stage, and reduces time limits for the review of AMDAL documents. It also advances the deadline for submitting environmental monitoring and management plans, so they can be reviewed and approved at the same time as the underlying AMDAL document, and calls for greater BAPEDAL involvement in the review of projects with multi-sectoral implications.

The laws and regulations of AMDAL in Indonesia which now exist are basically adequate. These have been developed over a decade and are now relatively comprehensive. However, the compliance and enforcement of AMDAL requirements are not sufficiently strict. In particular, the linkages between review findings and the permit and licensing conditions imposed on individual projects are tenuous (Smith and Van der Wansem 1995).

3   EA Process

In terms of process, AMDAL generally involves only four main steps of assessment – identify project features; identify environmental conditions; compare the project and environmental features in order to identify potential impacts; and evaluate the impacts and determine which should be mitigated and how. The detailed procedures for this are as follows (see Figure 12.1):

1. *Initiation.* The project proponent initiates the process by contacting the agency with authority for AMDAL.
2. *Initial project screening.* Government projects and non BKPM private projects are screened by the responsible government authority while BKPM (National Investment Board) projects are screened by an intersectoral team coordinated by BAPEDAL.
3. *Significance determination.* If a project is likely to result in significant impacts, a full EIA report and an accompanying environmental monitoring and management plan is required. If not, the project may proceed immediately if it meets standard design and operating requirements.
4. *Reporting.* If an EIA report is required, the Project Proponent should first draft Terms of Reference, for review and approval in conjunction with the reviewing commission, and then prepare the EIA report and submit it to the responsible AMDAL Commission. The EIA report must contain the environmental impact monitoring and management plans. Once the EIA and environmental impact monitoring and management plan documents have been submitted, the Commission has 45 days to decide whether: (a) to reject the project because the impacts are unacceptable, in which case the Project Proponent can revise or abandon the project; or (b) to allow the project to proceed. The Commission's approval can be conditional or unconditional.
5. *Final decision.* Final decisions on projects within the central bureaucracy are made by the sectoral Minister based on a recommendation from the appropriate central AMDAL Commission. At the provincial government level, final deci-

sions are made by the Governor based on a recommendation of the Regional or Provincial AMDAL Commission.

6. *Public involvement.* The public may be involved at any stage of the AMDAL process, at the discretion of the Commission. Comments can be submitted to the Commission before permit decisions are made. AMDAL regulations also require the authorized government agency to inform the public of activities requiring EIA documents and of its decisions.

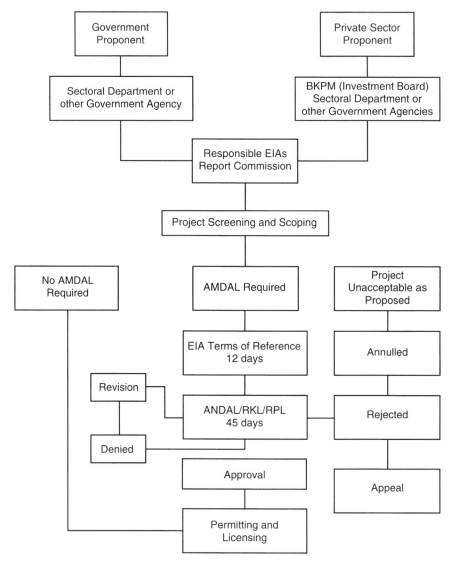

Figure 12.1   *AMDAL Process for New Projects in Indonesia*

In Indonesia, environmental impact monitoring and management plans are specified explicitly in AMDAL Regulations. These environmental impact monitoring and management plans are submitted to the appropriate AMDAL Commission for approval. In principle, no permits or licenses for constructing or operating a project are granted before approved environmental impact monitoring and management plans are in place. According to Government Regulation No. 51, environmental impact monitoring and management plans must now be submitted simultaneously with the EIA reports so that a single and consistent review can be conducted before the responsible AMDAL Commission approves or refuses projects, permits and licenses.

Although there has been a great expansion in the application of EIA in Indonesia its level of sophistication varies. For example, a post project review of the AMDAL study for the Saguling Dam and Involuntary Resettlement evaluated the problems which arose after the AMDAL study was undertaken (Nakayama 1998). The EIA methodology applied in this study succeeded in identifying the major environmental impacts caused by the project, from planning to operation. However, not all of the possible environmental problems which were later observed have been identified by the AMDAL. Moreover, some problems that were identified were not properly dealt with during the implementation of the project.

4 EA Institutions

BAPEDAL is the central level EIA agency that has been created to assume certain responsibilities for organizing and supervising the administration of the EIA system. In carrying out its function, BAPEDAL is expected to clarify EIA policies and practices, publish information relating to EIA, issue procedural and technical guidelines, resolve interagency coordination difficulties and disputes, provide technical assistance to project proponents and periodically audit all phases of the EIA process to make sure that it is thorough, fair and effective. BAPEDAL is also expected to maintain complete records of pending and completed EIA decisions and to provide reasonable notice of and access to such decision and materials.

Under the guidance of BAPEDAL, separate AMDAL Commissions have been established in 14 sectoral governmental departments and institutions. Each of the Commissions is chaired by a senior agency officiai and generally advance agency policies and priorities. Each central AMDAL Commission consists of several permanent members that are representative of functional institutions such as BAPEDAL, Department of Home Affairs, BKPM and BPN (National Land Agency). Each Commission also includes non-permanent members from related institutions and other organizations (e.g. NGOs, community leaders, scientists).

The movement toward decentralizing and delegating EIA implementation and decision making authority is evident in Indonesia. Some responsibility for EIA preparation, review and approval actions has already been decentralized. Meanwhile, the provincial Governor is required to establish regional AMDAL Commissions. The function of a regional commission is to evaluate AMDAL studies for activities funded from the regional budget, the national budget, private sector activities that require regional permits, and other activities under regional jurisdiction. In response to this development, regional and provincial institutions need to be strengthened and linked effectively to the central authorities.

Efforts to build regional and provincial capacity to conduct EIA need financial and technical support. As is similar in other developing countries, budgetary constraints inhibit institutional development. An additional constraint is the insufficient numbers of permanent members of commissions and the huge, diverse volume of issues with which they have to deal. No one is assigned to work full-time on AMDAL management, so these activities are additional to a team member's primary responsibilities. It is also unclear whether working on an AMDAL commission is considered relevant to the career prospects of commission members from government agencies. Therefore, AMDAL duties continue to be considered secondary business by permanent members of the commissions and are often delegated to lower ranking staff.

5   Conclusions

Since development activities will generally alter the environment, Indonesia uses AMDAL (EIA) to encourage sustainable development. The AMDAL process was established in law through Government Regulation No. 29 of 1986 which was promulgated under Law No. 4 of 1982. This regulation was revoked and replaced by Government Regulation No. 51 of 1993 and subsequently by several guidelines (BAPEDAL 1994). The existing laws and regulations are basically adequate but it is necessary to improve compliance and enforcement.

During 12 years' implementation of AMDAL, environmental awareness has increased among bureaucrats, industrialists and the community. However, the effectiveness of its implementation still needs to be increased. There is still insufficient understanding that AMDAL should be used as a planning tool to help decision makers balance competing economic, technical, social and environmental concerns. Too often, the agency responsible for approving the EIA study lacks the legal authority to modify project creation and design decisions or the political strength to improve the conditions of the project evaluation. More attention and effort are required in order to build regional, provincial and national capacities to conduct EIAs. It is also necessary to clarify the roles and responsibilities for organizing and supervising the administration of the EIA system (Collier 1997).

References

BAPEDAL (1994) *A Guideline on Environmental Impact Assessment (AMDAL) in Indonesia*, Jakarta, Indonesia
Collier, WL (1997) Environmental Management Institutions in Indonesia, *Working Paper No.5* Regional BAPEDAL Institutional Development Program, Jakarta, Indonesia
Nakayama, M (1998) Post-project review of environmental impact assessment for Saguling Dam for involuntary resettlement, *Water Resources Development* **14**: 217–229

Smith, DB and Van der Wansem, MVR (1995) *Strengthening EIA Capacity in Asia: Environmental Impact Assessment in the Philippines, Indonesia and Sri Lanka*, World Resources Institute, Washington DC

12.3   EA IN THE RUSSIAN FEDERATION

Aleg Cherp

1   Historical Overview and Legal Framework

The EA system in the Russian Federation has its roots in the so-called procedure of State Environmental Expert Reviews (SERs)[1] introduced in the USSR in the late 1980s in response to the growing awareness of domestic environmental problems and, to some degree, developments in Western EA. SERs are conceptually similar to traditional 'expertizas' (reviews or appraisals of plans and projects by expert committees) which used to serve as a co-ordination and control mechanism of the Soviet centrally planned economy. In order to facilitate the process of SER, the description of some environmental effects of an action and the mitigation measures were included in project documentation submitted to expert committees. Such description, an analogue of an Environmental Impact Statement (EIS), was termed OVOS[2]. Therefore, the Russian EA system is sometimes called the SER/OVOS system (Cherp and Lee 1997).

Though the foundations of the SER/OVOS system were laid in the USSR in the 1980s and even earlier, it was not until the 1990s that preventive environmental regulation based on SER became fully institutionalized in two laws:

- The Law on the Protection of Natural Environment (1991, amended in 1993)
- The Law on Environmental Expert Review (1995)

These acts establish a framework for Environmental Expert Review, defined as a procedure of reviewing plans for proposed activities with the purpose of 'establishing the conformity of the planned economic and other activity to ecological requirements and in determining whether or not an object of Environmental Expert Review may be realised' (Law on Environmental Expert Review 1995).

The main form of Environmental Expert Review is SER. The competent authority for SER is the State Committee for Environmental Protection of the Russian Federation (Goskomecologia). It is mandatory for most development projects and non-project actions. The outcome of the SER is the 'SER Resolution' which may be positive or negative. Another type of Environmental Expert Review is Public

---

[1] In Russian: 'gosudarstvennaya ekologicheskaya ekspertiza', also frequently translated as 'State Ecological Expertise', 'State Environmental Examination', or 'State Environmental Assessment'.
[2] The Russian abbreviation for 'assessment of impacts on the environment'.

Environmental Expert Review or PER which may be conducted by qualified NGOs and results in a non-binding Resolution.

The law stipulates that the documents submitted to SER must contain 'materials of assessment of the impact on the natural environment' (Article 14, Law on Environmental Expert Review 1995). The definition of what exactly constitutes these materials has been left to the competent authority which issued a number of secondary legislative acts regulating the content of documentation submitted to SER (for example, Ministry of Environment 1995, State Committee 1997b).

The requirement for developers to conduct an OVOS is contained in the 'Regulations on Assessment of Environmental Impacts (OVOS) in the Russian Federation' issued by the Ministry for Environmental Protection (a predecessor of Goskomecologia) in 1994. This is defined as 'a procedure for taking into account environmental requirements of the legislation of the Russian Federation in preparing and making decisions on social and economic development' (Ministry of Environment 1994). The OVOS Regulations contain only general requirements as to how developers must conduct the assessment of environmental impacts. They contain the principle of starting OVOS as early as possible in the project development cycle, considering alternatives, and conducting public hearings or organizing other suitable forms of public participation. However, they provide few details on the stages of OVOS, the ways in which to document its findings, and the usage of the latter in decision-making. Goskomecologia has announced its intention to revise these OVOS provisions and has issued a draft federal regulation on the Assessment of Impacts of Planned Economic and Other Activities on the Environment (State Committee 1998) which address some of the uncertainties of the previous Regulations.

The International Centre for Educational Systems (ICES), which is affiliated to the World Bank in Russia, has developed and published non-mandatory OVOS Guidelines that have been recommended by the Ministry of Environment as a training guide. The Guidelines describe four stages of the OVOS process: notification of intent, the production of a draft EIS, the production of an EIS and a public hearing. The stages of OVOS are related to the stages of the project development cycle and to the stages of project authorization by the authorities (including SER). The ICES Guidelines are compatible with the World Bank EA procedures and have been used by a number of investors using Bank loans. However, they have never been officially endorsed and their usage by developers and authorities, especially in the provinces, seems to be quite limited.

2   Stages in the SER/OVOS Process

Screening

SER is mandatory for all project-level developments, independent of their size and nature, and for most policies, plans and programmes, including draft laws (Law on Environmental Expert Review 1995). Such a wide definition of the scope of SER leads to a large number of Reviews being conducted annually: more than 90 000 in

1996 (State Committee 1997a). In practice, competent authorities exempt certain minor developments (for example, construction of individual buildings in conformity with Master Plans already approved by SER) from SERs or prescribe a 'simplified' SER procedure. Such policies are adopted either at the discretion of regional Environmental Protection Committees, or regulated by regional legislation[1].

Unlike SER, the OVOS procedure is required only for certain types of projects listed in Executive Order 222 (Ministry of Environment 1994). The OVOS screening list is slightly broader than the list of developments subject to EIA in a transboundary context according to the UNECE Espoo Convention (UNECE 1991). In addition, provincial and federal governments are given the discretion to initiate an OVOS procedure for other types of environmentally significant projects. There are no statistics on the number of OVOSs conducted in accordance with Order 222. The intended application of a universal environmental assessment procedure to all activities which are subject to SER has been formally stated in the draft 'Main Principles of Environmental Assessment' (State Committee 1998).

Scoping

The existing Russian legislation does not require scoping in either preparing the documents for SER or in the OVOS procedure. In practice, in determining the scope of an assessment, developers are guided by the 'Construction Norms and Rules' and other legal standards, and the requirements to provide the necessary documented information to SER, as required, for example, by the Ministry of Environment (1995). A number of guidelines developed by ministries and larger companies may also be used in determining the scope of EA of particular developments.

The OVOS Guidelines (ICES 1996) suggest obtaining scoping recommendations from local and environmental authorities after submitting a preliminary EIS. An approach presented in the current draft of the 'Main Principles of Environmental Assessment' (State Committee 1998) mentions the scoping stage during which the developer defines the terms of reference for the consultant who carries out the EA, no other parties being involved. In all of the above mentioned approaches to scoping, the involvement of the public and the production of a scoping report is not required.

EIS Preparation

The existing OVOS Regulations (Ministry of Environment 1994) do not contain a specific requirement to prepare an EIS or its analogue. The Law on Environmental Expert Review refers to 'environmental assessment materials', which are to be

---

[1] For example, Goskomecologia's order regulating the SER procedure (State Committee 1997b) establishes a different duration of SER for 'simple', 'medium' and 'complex' developments. Nizhny Novgorod regional 'Regulations on State Environmental Expert Review' (Legislature of Nizhny Novgorod province 1997) distinguishes between 'simple' and 'complex' activities which are subject to SER.

submitted to SER. This has been further elaborated by the Ministry of Environment in its *Instruction on Ecological Substantiation* of Economic and Other Activities (Ministry of Environment 1995). 'Ecological substantiation' is not a separate document, but a 'set of arguments (evidences) and scientific forecasts, to evaluate the degree of environmental danger of the proposed activity' (Article 2). The 1995 Instruction requires the 'ecological substantiation' to contain a description of certain parameters of the project (for example, inventory of projected discharges, emissions and solid waste) and of the affected environment but there is no explicit requirement to analyse the possible impacts (especially, indirect and cumulative impacts).

The OVOS Guidelines of ICES (ICES 1996) recommend the preparation of draft and full EISs (ZVOS) (in addition to 'OVOS' and 'Nature protection' volumes of the project documentation) which should be used in SER and other project authorization proceedings as well as in public hearings. The draft *Main Principles of Environmental Assessment* (State Committee 1998) also proposes the production of a separate EA report: however, it is not suggested that it be used in project authorization procedures nor that it be available to the public.

Consultation and Public Participation

The public can take a limited part in both SER and OVOS procedures as well as organize Public Environmental Expert Reviews (PERs).

During the SER procedure, the public has the right to send its 'well-reasoned proposals' to SERDs and the authorities are required to explain how these proposals were considered during the SER (Law on Environmental Expert Review 1995). However, since it is not obligatory for assessment information to be made available to the public at this stage, it is not clear how people are supposed to prepare their comments. At the end of the SER process, the local authorities are required to inform the public about its findings.

The OVOS Regulations 1994 also contain certain public participation provisions. The developer is required to conduct public hearings or public discussion in the mass media relating to the proposed activity where it is not 'a specially designated object, the information about which is confidential' (Ministry of Environment 1994). Unfortunately, the definition of public hearing and public discussion is not given in the Regulations. Since there is little experience of public participation, developers are left uncertain regarding their responsibilities in this area.

The legislation also provides for a separate non-mandatory process of PER that can be organized by registered NGOs. PER must be conducted at the same time as SER. PER is similar in concept to SER, but the crucial differences are that PER experts are appointed and paid by public associations, their right of access to project documentation is not guaranteed, and PER Resolutions are not legally binding.

According to the practical experience of dozens of recently organized PERs, the most significant constraint of this mechanism of public participation is the inability to start a PER if 'its object contains commercial or other secrets' (Article 21, Law on Environmental Expert Review 1995). Since *all* project documents, not just environmental assessment materials, are subject to PER, they typically contain some 'secrets', especially as the concept of a commercial secret is not clearly defined in Russian law (Khotuleva *et al.* 1996).

A precondition for a positive Resolution is endorsements of the activity by public health, industrial safety, and several other statutory authorities. Obtaining these endorsements is the main mechanism of consultation within the Russian EA system. The Regulations do not specify at what stage of the pre-SER environmental assessment process these authorities should be consulted.

Review, Decision-making and Post-project Analysis

After preparing the project documentation, the developer submits it to the SER competent authority, i.e. Goskomecologia, for developments of federal importance, or a provincial Environmental Protection Committee for other activities. SER is a procedure for reviewing the project (or other planning) documentation, checking (a) the comprehensiveness and reliability of the information presented, and (b) the environmental acceptability of the proposed activity. The review can be conducted either by the staff of the competent authority (normally for minor developments) or by an independent panel of experts appointed by the competent authority. Due to the large number of SERs which are conducted, most of them seem to be limited to checking project compliance with environmental norms and standards. For selected larger developments, SERs may involve elements of more sophisticated expert assessment of the significance of the expected environmental impacts of the proposed activity.

The outcome of a SER is a legally binding Resolution. If the Resolution is negative (which happens in 20–30% of all cases undergoing SER for the first time), the action cannot be implemented. A positive Resolution may contain recommendations relating to the environmental conditions to be met in project implementation and for environmental monitoring. These conditions are the basis for inspections conducted by Environmental Protection Committees as part of post-project analysis.

3 Environmental Assessment of Strategic and Other 'Non-project' Actions

The OVOS procedure, as defined in the OVOS Regulations (Ministry of Environment 1994) and interpreted in the OVOS Guidelines of ICES (ICES 1996), is only required for project-level developments. However, SER is also mandatory for many strategic developments: virtually all land use and sector plans, federal and regional programmes and policies, new products and technologies, environmental protection rules and standards and draft legislation. Consequently, the assessment of the environmental impacts of these activities is needed prior to SER. The information ('ecological substantiation') to be prepared for SER of certain strategic actions, such as land-use plans, sectoral and regional development schemes, is regulated by the 1995 Instruction (Ministry of Environment 1995). SER of strategic actions is similar in legal status to SER of projects, but for draft laws the SER Resolution is not legally binding.

In practice, SER and its associated pre-SER environmental assessment are consistently applied to physical plans as well as to certain federal programmes. The extent of the practical application of SER to other types of strategic actions is unclear due to the absence of well-defined regulatory frameworks and methodological and procedural guidelines. As far as the assessment of strategic activities is concerned, the SER approach which is largely based on checking the compliance with norms and standards is not entirely applicable. It is also not clear whether Goskomecologia, when conducting SERs, has a mandate to issue binding SER Resolutions regarding these activities. All these factors weaken the practice of applying SER to strategic actions.

4   Conclusions and Suggested Improvements

The Russian EA system has its roots in the Soviet system of centrally controlled planning and was established during a time of social and economic transition. In the changing circumstances of transition there are, in the author's opinion, three key areas where the Russian system should be significantly improved:

- Strengthen pre-SER environmental assessment
- Adopt a more differentiated approach to environmental assessment for different types of activities
- Improve the transparency of the EA procedure

These matters are now widely debated and are partially reflected in new draft federal regulations (for example, State Committee 1998) and in some regulations at the regional level (for example, Legislature of the Nizhny Novgorod region). Additionally, some non-regulatory (capacity building) initiatives that could significantly improve the quality of EA practice are also needed.

Status of Pre-SER Environmental Assessment

The current regulatory and institutional system regards SER as the main element of the EA process and considers environmental assessment conducted by the developer as a preparatory stage for SER. The pre-SER stages of the process are barely mentioned in the legislation and do not have strict procedural controls. This results in the production of documents of lower visibility and status than the SER Resolution. The negative consequences of such an approach are:

- There are no statutory incentives for the developer to start EA early in the project design stage in order to identify the most environmentally sound options of the design and implementation of the planned activity
- There is little transparency in the pre-SER process and no possibility for the affected parties to influence its scope, content and outcome; additionally the

interested parties are denied access to the pre-SER findings because no separate environmental assessment document is produced at this stage
- Much for the same reason, the advancement of pre-SER practice is hindered by the lack of accessible registries of pre-SER (OVOS) reports
- Capacity building of the actors of the key pre-SER stages of EA assessment (developers, the public, local authorities etc.) is restricted by the unbalanced distribution of resources, most of which go towards strengthening the SER component

These deficiencies can be remedied by imposing procedural checks on pre-SER environmental assessments conducted by developers, introducing a requirement to produce a separate pre-SER EA report (which is available to the public, which can be used in diverse decision making procedures, and which is placed in an accessible central depository to enable learning from past experience to be diffused).

Differentiated Approaches to Environmental Assessment

The lack of screening provisions, which has resulted in very large numbers of projects being submitted to similar levels of assessment, has prevented improvements in regulatory standards and in compliance with SER/OVOS requirements. Developing and applying EA procedures of different degrees of complexity, according to the environmental significance of the proposed action, would be a logical approach to solving this problem.

Additionally, the proponents of 'non-project' activities, including strategic policies, plans and programmes, require more detailed guidance on the different levels of detail and forms of environmental assessments for these.

Transparency and Public Participation

The existing provisions for public participation in the Russian SER/OVOS system are rudimentary. They provide for very late and limited involvement of the public, they lack clarity regarding the responsibilities of developers and authorities, and they allow broad discretionary exclusions of planned activities from public scrutiny. If public participation happens in practice, it is often only due to the persistence of stronger environmental NGOs. PERs can be a tool for consolidating and expressing the opinions of special interest groups with significant resources, but they can hardly serve the purpose of informing and involving the wider public, including local communities, in environmental decision making.

Public participation in the Russian EA system can be significantly strengthened by:

- Introducing a requirement for pre-SER EA documents to be available to the public as well as SER Resolutions
- Mandatory notification of the affected public of the start of the pre-SER EA process
- Mandatory public hearings for certain types of developments

Capacity Building and Other Activities

Significant improvements in EA practice will depend upon capacity building for all actors in the EA process. The main priorities are:

- Producing and disseminating guidelines for developers on methods for carrying out their environmental assessments, and for competent authorities on the ways of checking the quality of these
- Conducting workshops and training seminars for wide audiences, commencing with the environmental authorities. These should introduce the concepts and approaches of environmental assessment and identify the opportunities for utilizing these in the SER/OVOS system
- Promoting monitoring and performance evaluation activities and facilitating learning from experience in other ways, for example, by maintaining open depositories of EISs. Independent researchers should play a more important role in these activities
- Raising the awareness of local authorities, the general public and NGOs of the possibilities and methods of participation in the SER and OVOS procedures

In the present economic situation in Russia, many of these activities are hardly possible without assistance from the developed countries. However, so far, Western assistance in the field of EA has been relatively insignificant, unfocused and uncoordinated. In the future, international and bi-lateral development agencies should more closely target their activities to the specific needs and audiences of the region.

To conclude, the Russian EA system contains most of the necessary elements to become an effective tool of environmental protection. The current deficiencies of the system can be remedied by establishing stricter procedural controls and greater transparency of the pre-SER stages of environmental assessment, through introducing differentiated requirements for EA of activities of different environmental significance and through implementing a number of institutional and capacity-building initiatives.

References

Cherp, O and Lee, N (1997) Evolution of SER and OVOS in the Russian Federation, *EIA Review* **17**: 177–204

Cherp, O and Khotuleva, M (1999) Environmental assessment of a waste incineration plant in the Moscow region of Russia, in *Case Studies of Environmental Assessment in Countries in Transition*, Lee, N (ed.), Central European University, Budapest (in press)

International Centre for Educational Systems (ICES), EIA Department (1996) *Guidelines on Conducting Assessment of Impacts on the Environment in Developing Substantiation for Investments in Feasibility Studies and Projects of Construction, Re-Construction, Expansion, Modification, Conservation or Decommissioning of Economic and other Facilities and Complexes*, ICES, Moscow

Khotuleva, M, Cherp, O and Vinichenko, V (1996) *How to Organise a Public Environmental Review: A Guide to NGOs*, ECOLOGIA-ECOLINE, Ecoline, Moscow (in Russian)

Law of the Russian Federation on Environmental Review (1995) 23.11.95, No. 174-FZ

Law of the Russian Federation on the Protection of Natural Environment (1991) 19.12.91, No. 2060–1 with amendments of 02.06.93, No. 5076–1

Lee, N (1995) Environmental assessment in the European Union: a tenth anniversary, *Project Appraisal* **10**: 77–90

Ministry of Environmental Protection and Natural Resources of the Russian Federation (1994) Order No. 222 of 18.06.94. *On Introducing the Regulations on the Assessment of Environmental Impacts*, Moscow, Russia (in Russian)

Ministry of Environmental Protection and Natural Resources of the Russian Federation (1995) Instruction on Environmental Substantiation of Economic and Other Activities, Attachment to the Order No. 539 of 29.12.1995 Moscow, Russia (in Russian), unpublished

Nizhny Novgorod Oblast Legislative Assembly (1997) Decree of 25.03.97 N 44 On Introducing the Regulations on the State Environmental Review in the Nizhny Novgorod oblast

State Committee of the Russian Federation for Environmental Protection (1997a) *The National Report on the State of the Environment of the Russian Federation in 1996*, The Centre for International Projects, Moscow

State Committee of the Russian Federation for Environmental Protection (1997b) Order on Conducting State Environmental Review No. 280 of 17.7.97

State Committee of the Russian Federation for Environmental Protection (1998) The Main Principles of the Assessments of Impacts of Planned Economic and Other Activities on the Environment (draft), http://cci.glasnet.ru/mc/eia/eia_draft.html

UNECE (1991) Convention on Environmental Impact Assessment in a Transboundary Context, UNECE, Geneva

13

Country Studies of EA in Nepal, Jordan and Zimbabwe

13.1   EIA IN NEPAL

Ram B. Khadka and *Batu K. Uprety*

1   Introduction

Nepal, which is situated in the central Himalayas between the arid Tibetan plateau in the north and the fertile Gangetic plain in the south, is a small mountainous country with an area of 147 181 km². It is characterized by diverse physiographic zones, contrasting climates and altitudinal variations ranging from 75 to 8848 m (Mount Everest). Topographically, the country is broadly divided into three distinct ecological regions. They are: mountains, hills and the Terai occupying 35, 42 and 23% of the total land area, respectively. Only 8% of the total population inhabit the mountain region because of its rugged terrain and its harsh climatic conditions. The population density in the hills is high, as 45% of the total population live there. The Terai is most fertile and contains 47% of the population. Of the total population of Nepal, 90% live in rural areas (CBS 1997).

In general, the mountains and hills have steep slopes and are geologically fragile and highly erodable. The large number of deep valleys and the vertical extension of the Nepalese Himalayas have contributed to the formation of many isolated localities in the hills and mountains. The Terai, being the most fertile land formed with rich alluvial materials, have been utilized for extensive agriculture. The highland interacts with the lowland and any kind of ecological disturbance that occurs in the highland also manifests itself in a series of environmental effects downstream in the form of floods, landslides, erosion and desertification. Therefore, the development activities that are being implemented in the mountain and hills give rise to

Environmental Assessment in Developing and Transitional Countries. Edited by N. Lee and C. George.
© 2000 John Wiley & Sons, Ltd.

cumulative and multiple environmental effects both at their origin and downstream (HMG 1998).

2   Environmental Issues and Problems

The environmental problems of Nepal are of both an urban and rural character. The urban areas mostly suffer from unplanned and haphazard development patterns. Inadequate solid waste management systems, gaseous emission and the disposal of untreated sewage to the land and in water bodies has led to the contamination of air, water and land and increasing levels of noise. These are some of the emerging environmental problems in the urban areas of Nepal.

Additionally, urban pollution problems have been aggravated because of the absence of a land use plan or zoning regulations to guide development and to provide some control measures for the unprecedented growth of urban settlements in Nepal. Poverty is often blamed as the root cause of the growth of urban slums; however, this is not true in all cases. Affluence can be equally blamed for degrading urban environmental quality. Whatever may be the underlying cause, the intensity of airborne and water-borne diseases is increasing and thousands of lives are being lost or severely affected each year in Nepal as a consequence (EPC 1993).

In rural Nepal, land degradation, forest exploitation and sanitation are the major environmental concerns. Soil erosion, fertility decline, sedimentation and frequent floods have degraded the productive capacity of the land used in rural agriculture. Soil loss due to the cultivation of steep slopes and marginal land, overgrazing, deforestation and population pressure on agricultural land have led to frequent floods in the plains. These are the major causes of land degradation which have considerably affected agriculture productivity in recent years. This is a major concern for the subsistence of people inhabiting the rural areas of Nepal.

3   Initial Development of an EIA System in Nepal

The problems outlined above and the increasing pace of development have prompted the formulation of an EIA system in Nepal in order to integrate environmental considerations into development planning and implementation in the country. His Majesty's Government of Nepal introduced an environmental policy for the first time in its sixth Five Year Plan (1980–1985) and this was elaborated in its seventh Five Year Plan (1985–1990) (HMG 1985). During the period of the seventh FiveYear Plan, the government endorsed a National Conservation Strategy (NCS), within which the development of an EIA system for Nepal was a major component. Its implementation then started in 1989 and was later supported by the eighth Five Year Plan (1992–1997) which enunciated a separate policy on Environment and Resource Conservation as a part of national policy (HMG 1992, 1997a).

Although, the implementation of the National Conservation Strategy formalized EIA application through the involvement of government agencies, the concept of EIA was not entirely new to Nepal at that time. A minority of projects were subjected to EIA in the late 1970s but all of these were commissioned primarily in response to donor conditionality. In the case of the NCS, the integration of environmental considerations into the formulation of central and local development projects was the main objective.

The development of EIA guidelines for different sectors was given priority and fully supported by the eighth Five Year Plan. The Plan set a target to develop EIA guidelines for such large scale development projects as road construction, hydropower, irrigation, housing, drinking water supply and sewerage projects. Accordingly, HMG launched a separate EIA programme in collaboration with IUCN (The World Conservation Union). This programme has been a vehicle for creating awareness of the need for environmental assessment and for developing an EIA system in Nepal. The programme was designed, using a learning-by-doing approach, and trained a number of government officials, NGOs members and individuals from private sector agencies, using EIA training manuals developed for this purpose (Khadka and Tuladhar 1996; Khadka *et al.* 1996; Khadka 1997).

The eighth Five Year Plan has been important in developing and institutionalizing an EIA system in Nepal's development planning and administration. The government developed and approved umbrella national EIA guidelines in 1993 (HMG 1993). Separate guidelines for the forestry (Box 13.1) and industry sectors were developed and endorsed by the government in 1995 (HMG 1995a,b). Other sector specific guidelines have been developed and are being officially endorsed; others are in preparation (see Table 13.1).

All of these guidelines were developed using a participatory approach. They were drafted in workshops involving EIA stakeholders as government officials, representatives from NGOs and private sector organizations, corporate bodies, environmental consultancies and individual professionals. The guidelines were tested to validate their applicability in field conditions and tailored to suit local conditions. The use of the participatory process in developing EIA guidelines promoted wider stakeholder participation, facilitated their endogenous evolution and fostered a multi-disciplinary approach (Khadka and Tuladhar 1996)

Box 13.1   EIA in the Forestry Sector

About 42% of the total area of the country is covered fully or partially by forest and is managed by the Department of Forest. Most development and infrastructure projects are being implemented within forest areas or pass through them. Therefore, substantial areas of forest regions have been destroyed or degraded in the past in the process of implementing development projects. In order to protect the remaining forests, the Government of Nepal, through its Forestry Sector EIA Guidelines 1995 and Environmental Protection Regulation 1997, has made provision for carrying out an EIA for those projects which require felling of forest trees in areas of more than 5 ha. Felling of trees in areas of less than 5 ha requires an IEE.

Extraction of stones from a quarry, implementation of leasehold and private forestry activities and construction of hotels and resorts in the forest area require the application of environmental assessment provisions. Implementation of linear projects that pass through the forest area also require the application of environmental assessment. Such IEE and EIA reports should be approved by the Ministry of Forest and Soil Conservation and the Ministry of Population and Environment respectively (HMG 1996).

Table 13.1   *EIA Guidelines for Nepal*

Guidelines	Status of guidelines
National EIA Guidelines (1993)	Endorsed and currently in use
EIA Guidelines for the forestry sector (1995)	Endorsed and currently in use
EIA Guidelines in the industry sector (1995)	Endorsed and currently in use
EIA Guidelines for water resources (1996)	In the process of endorsement
EIA Guidelines for the road sector (1996)	In the process of endorsement
EIA Guidelines for the mining sector (1995)	In the process of endorsement
EIA Guidelines for urban development (1996)	In the process of endorsement
EIA Guidelines for the tourism sector (1996)	Drafted
EIA Guidelines for agriculture (1997)	Drafted
EIA Guidelines for landfill site selection (1997)	Drafted

4   Internalization of the EIA System

The sector specific EIA guidelines were developed under the broad framework of National EIA guidelines, 1993. Recently (see Section 5 below), the application of EIA guidelines has been made effective through the enforcement of environmental legislation. Prior to the legal regime on EIA, all large sized projects had to be submitted to the Environmental and Resource Conservation Division (ERCD) of the National Planning Commission for authorization. ERCD also had the responsibility to examine the project submitted whether or not it had undergone an EIA according to the national and sector-specific EIA guidelines. In this way, even in the absence of umbrella legislation, all large sized development projects proposed from the public sector, and projects proposed from the private sector which required authorization from a concerned government agency, have undergone an EIA utilising the national and sector specific EIA guidelines.

After four years of successful implementation of the EIA guidelines, and experience and benefits derived from this, the government has introduced several provisions to legally internalize EIA in the decision making process. EIA has now been made mandatory for prescribed projects and activities through the provision of the Environmental Protection Act 1996 and the Environmental Protection Regulation 1997. Two illustrations of the application of an existing EIA provision are provided in Boxes 13.1 and 13.2.

5   Environmental Legislation

Nepal's Environmental Protection Act (1996) obliges the proponent to carry out either an Initial Environmental Examination (IEE) or an Environmental Impact

---

Box 13.2   EIA in Bardibas–Jaleshwor Road Construction Project

Over the last 50 years, 11 000 km of roads have been constructed, of which 32% has been blacktopped. Road construction in Nepal's mountain area is difficult because of its rugged terrain and fragile geology. Most of the hill road is destroyed each year from landslides, river bank cutting and inadequate environmental considerations. To avoid investments of large amounts of capital for maintenance and rehabilitation of roads, and to make future road projects environmentally sound and sustainable, the Government of Nepal has started a system of integrating environmental impact assessment in the road construction projects according to Environmental Rules 1997, and the draft Road Sector EIA Guidelines 1995. One example of this was the construction of a 142 km North–South road (Bardibas–Jaleshwor) in central Nepal in 1997. The EIA was conducted of the proposed scheme and identified both positive and negative impacts, and proposed benefit maximizing measures and mitigation of adverse impacts.

The mitigation prescriptions were recommended to minimize the adverse impacts identified, and proposed bioengineering treatments, road side plantation and management. The study includes mitigation management responsibilities, the costs of the mitigation measures, and the monitoring and audit requirements. The project is now under construction and the environmental management provisions, as recommended by the EIA, have been incorporated into project construction. The final environmental audit has to take place after one year upon the completion of project construction and will indicate the effectiveness of the environmental provisions in the project.

Assessment (EIA) for the prescribed projects and programmes. For example, IEE is required for the establishment and/or expansion of protected areas, construction of 1–5 MW of electricity project, construction of district roads etc. Similarly, EIA is required for a forest management plan, national highway, electricity generating projects exceeding 5 MW etc. According to the Act the proponent is obliged to submit a proposal requiring an IEE for approval by the concerned agencies, or to the Ministry of Population and Environment (MOPE) in the case of EIA. Before the approval of the project MOPE must publish a notice relating to the availability of the EIA report for public comment and review. It may also form a technical committee to review the EIA report (HMG 1996).

Any proponent who implements a prescribed project without the approval of its IEE/EIA, or without compliance with any conditions imposed during proposal approval, shall be fined up to Rs 0.1 million (1 US$ = Rs 63). The designated officer from MOPE may also issue an order stopping project construction activities immediately. The Act empowers the environmental inspector to oversee and review project implementation and examine whether or not the proponent is complying with the conditions approved in the process of project authorization. If the project proponent is not satisfied with the decision made by the designated officer or environmental inspector, he/she has the right to appeal to the Appellate Court within 35 days from the date of the decision made or order issued (HMG 1996).

The Environmental Protection Regulation 1997 elaborates the provisions for the application of IEE/EIA. The proponent wishing to operate a 'prescribed project' in any given area is obliged to issue a public notice as a first step of scoping, describing the nature of the proposed project to be implemented in the area concerned. The proponent shall invite any expression of concern, within 15 days as per EPR 1997 (amendment 1999), from people in the area where the project is planned to be implemented. The project proponent shall collect and compile these responses within a scoping report which is to be submitted to the concerned government agency for approval. The development of the Terms of Reference (TOR) for further EIA studies should be based on the approved scoping report. The scoping report and the Terms of Reference form the basis for developing the IEE/EIA Report for the particular project in question. This is submitted to the concerned Ministry (in the case of the IEE) and to MOPE (in the case of the EIA) for approval.

Before submitting the IEE/EIA report to the concerned Ministry or MOPE the proponent has to make it available to the Village Development Committee and the District Development Committee and in schools and a public place where the project is likely to be implemented, for public review. A public hearing should be made in the project area to collect suggestions (HMG 1998). All the comments and suggestions provided by the concerned public should be compiled and incorporated into the final report on IEE/EIA (HMG 1997b).

According to the legal provision, the concerned Ministry should approve the final IEE report within 30 days after its submission, and the Ministry of Population and Environment should approve the EIA report within 90 days following submission. However, in the case of EIA, MOPE should make the final report available to the public for review and the time period given for public review in this case is 30 days. MOPE may also form an EIA Review Committee of experts which may also include representation from the communities likely to be affected by the project.

Implementation of the mitigation measures prescribed in the IEE/EIA and recommended by the project authorizing agencies, and other conditions imposed by the approving agencies, are the responsibility of the project proponent. However, EPR 1997 also empowers the concerned Ministry to monitor the implementation of the compliance requirement by the government agencies (Box 13.3). MOPE is also responsible for environmental auditing after the completion of the project (HMG 1997b).

Box 13.3   Environmental Monitoring of Hydropower Projects

In the past, EIA reports were used for project approval by the licensing authority. Post EIA activities were not mandatory. However, arising from the Environmental Protection Regulation 1997, environmental monitoring and auditing activities of project implementation have become a mandatory requirement. The construction of three major hydropower generation projects are currently being monitored from an environmental point of view. They are: (a) Kaligandaki; (b) Khimti; and (c) Upper Bhotekoshi.

Project (a) belongs to the government but (b) and (c) are being constructed by private sector companies, and the construction of (a) and (b) are monitored by the proponents themselves. However, project (c) is being monitored by a third party.

The upper Bhotekoshi hydropower project is being monitored by IUCN Nepal. Prior to the construction of the project, a pre-audit study was undertaken which identified major areas for compliance, impacts and baseline monitoring. Indicators for each of these were developed and were being measured during the construction period.

The monitoring process will be continued up to January 2000. After construction is completed, a post audit study is proposed to be undertaken, which will examine the effectiveness of the integration of EIA into project construction.

6 Conclusions

The historical evolution of the EIA system in Nepal is quite interesting. The national policy was first translated into EIA policy guidelines which were to be used as required by the National Planning Commission (NPC) of the government. The NPC, being an apex advisory body within government, issued a notice to all sector specific ministries and government departments to adopt EIA guidelines for application to development projects in their respective sectors. The NPC also examined all projects submitted for government approval and, if it found there was no integration of environmental considerations within the proposal as required in the respective EIA guidelines, approval was not granted. This mechanism worked well, even in the absence of umbrella environmental legislation, and provided an opportunity to test the EIA guidelines and to tailor the system to local conditions. Later, it was accepted that the EIA system could work well in the country, and that it was an important tool to make development projects environmentally sound and sustainable. The Ministry of Population and Environment, established in 1995, re-emphasized EIA process development and environmental protection legislation, enforced since June 1997, provided the necessary legislative back-up for internalizing EIA in Nepal's development process.

Much has been done so far in establishing and implementing an EIA system in Nepal. However, many challenges remain to be met before full, systematic implementation of EIA can be realized. The most important areas requiring further action in the near future are:

- The development of environmental management procedures, keyed to the organizational and institutional framework of different ministries, is required to operationalize the sector specific EIA guidelines (e.g EIA report review criteria, checklist for reviewing on-going and operational projects). While the sector specific EIA guidelines, produced to date, provide essential technical guidance for EIA studies, implementation of the guidelines must be tied to the project planning process of the respective departments, agencies and organizations
- Different kinds of activities have helped in raising awareness of the importance of EIA at the central level. However, it is equally important to have the same level of awareness at the local and district level. There is an urgent need to initiate awareness raising activities and EIA training at the local level of administration. EIA guidelines specific to small scale initiatives should be developed for IEE/EIA application at the local level.
- Human resources for EIA implementation, both at the professional and managerial levels, are critically inadequate in the country. There is an urgent need to develop a long term human resource development plan for key departments, agencies and organizations involved in the EIA system. Arrangements with teaching institutions, both national and foreign, are required, in order to produce an adequate number of trained personnel to meet present requirements and future needs
- Public participation is the heart of EIA. An effective system of public participation, which also covers gender issues, public hearings and public review procedures, should be established

• A system for using fiscal and other economic incentives to accompany EIA implementation should be initiated. Combining the use of EIA and economic evaluation techniques could do much to support the integration of environmental considerations into Nepal's development planning

References

Central Bureau of Statistics (CBS) (1997) *Statistical Year Book of Nepal*, Central Bureau of Statistics, Kathmandu

EPC (1993) *Nepal Environmental Policy and Action Plan (NEPAP)*, Environmental Protection Council, Kathmandu

HMG (1985) *The Seventh Five Year Plan*, National Planning Commission, Kathmandu

HMG (1988) *Building on Success: The National Conservation Strategy (NCS) for Nepal*, National Planning Commission/IUCN, Kathmandu

HMG (1992) *The Eighth Five Year Plan*, National Planning Commission, Kathmandu

HMG (1993) *National EIA Guidelines*, HMG Press, Kathmandu

HMG (1995a) *EIA Guidelines of Forestry Sector*, HMG Press, Kathmandu

HMG (1995b) *EIA Guidelines for Industry Sector*, HMG Press, Kathmandu

HMG (1996) *Environment Protection Act, 1996* Ministry of Population and Environment (MOPE), Kathmandu

HMG (1997a) *Consolidated Report, National Conservation Strategy Implementation Project*, National Planning Commission/IUCN, Kathmandu

HMG (1997b) *Environment Protection Rules, 1997*, Ministry of Population and Environment, Kathmandu

HMG (1998) *Environment Protection Rules, 1997 (amendment 1999)* HMG Press, Kathmandu

Khadka, RB and Tuladhar, B (1996) Developing an environmental impact assessment system in Nepal: a model that ensures the involvement of people, *Impact Assessment* **14**: 235–248

Khadka, RB (1997) *EIA Manual for Senior Government Officials*, IUCN, Nepal

Khadka, RB, Neam, P and Bisset, R (1996) *EIA Training Manual for Professionals and Managers*, AREAP, IUCN, Nepal

13.2   EIA IN JORDAN

Mahmoud Al-Khoshman

1   Introduction

This section provides an overview of the existing situation of Environmental Impact Assessment (EIA) in the Hashemite Kingdom of Jordan. It contains a brief description of the current status of EIA guidelines and procedures, the main difficulties being encountered in EIA implementation in the country, proposed requirements for a successful EIA system, and some recent developments.

Despite many trials and proposals by national and international institutions over the last decade, there is currently no legal requirement for EIA in Jordanian legislation nor any EIA guidelines or procedures, although an EIA by-law is in preparation. The lack of a specific legal requirement is a fundamental weakness. The preparation of a national EIA regulation and of implementation guidelines is a priority, together with the development of a framework for other necessary environmental legislation (pollution control, environmental quality, protected areas etc.)

As an encouraging preliminary step, the Environment Protection Act, 1995 (referred to hereafter as 'the Act') provides a general directive regarding EIA. Article 15 of the Act states that: 'the General Corporation for Environmental Protection shall establish national EIA procedures and guidelines for projects to ensure environmentally sound developments and achieve sustainable development.'

2 Aqaba Region Authority EIA Operational Directive, 1995

In October 1993, the World Bank issued Report No. 12244 JO entitled *Gulf of Aqaba Environmental Action Plan – JORDAN* (GAEAP). The report proposed a plan to prevent further environmental degradation of coastal areas, coral reefs and marine ecology in Jordan's Aqaba region while promoting sustainable economic development. An essential feature of this plan is the preparation of a comprehensive regulatory and institutional framework for the management of Jordanian marine waters and coastal areas.

Component (A) of GAEAP calls for the creation of a legislative and regulatory framework for the environmental protection of the Gulf of Aqaba, including the following: (a) identification and definition of air and water polluting substances; (b) development of air and water quality standards and regulations; (c) establishment of regulations pertaining to zoning and siting of all developments, and activities such as tourism and fisheries; and (d) development of a system of environmental impact assessment (EIA) review and permit procedures. In order to execute this, the World Bank and the Aqaba Region Authority (ARA) have sponsored the Jordan Marine Pollution Action Plan (PPA).

The Aqaba Region Authority EIA Operational Directive of 1995 (ARA EIA OD '95) has been prepared as part of the PPA, to provide the ARA with an effective planning tool to ensure that proposed projects in the Aqaba Region take environmental considerations into account. Under this proposed directive, EIA would be required of all projects that may have significant effects on the environment, natural resources and community by virtue of their location, resulting activities, nature or size. Public sector as well as private sector projects would be subject to its provisions.

The Directive aims to make environmental planning an integral part of the planning system of the Aqaba Region Authority. It would provide the ARA Environment Unit with a powerful instrument which should enable it to screen proposed projects and developments and ensure that potentially adverse environmental impacts are assessed and appropriate mitigation measures are integrated into design, construction, operation and decommissioning plans (Al-Khoshman 1995)

3 International EIA procedures

In the absence of national or regional EIA procedures in Jordan, the EIA studies which have been prepared have followed the procedures of international organizations and bilateral donors. This sub-section reviews the procedures of three organizations whose procedures have been used in the environmental assessment of some projects in Jordan.

The World Bank Environmental Assessment (EA) Operational Directive (OD 4.01)

This directive was first introduced as a formal directive in 1989 (as OD 4.00), then was enhanced and re-introduced in September, 1991 (as OD 4.01) and, most recently in January 1999 was converted into Operational Policy (OP) 4.01. This directive contains the World Bank EA procedures that apply to projects financed by the Bank.

The preparation of environmental assessments for projects proposed for World Bank support is the responsibility of the co-operating Government. They are expected to address the applicable national environmental policies and legislation. In the case of co-financed projects, they should also address the environmental procedures of the co-operating financial institutions and donor organizations.

In Jordan, the World Bank EA procedures have been followed in various Bank financed projects including the Red Sea–Dead Sea Canal; Aqaba Thermal Power Station Stage-II; and the Adasiya Dam on the Yarmouk River. Boxes 13.4 and 13.5 contain brief outlines of two of these EIAs.

The European Investment Bank (EIB)

The EIB uses a version of the European Union's Environmental Assessment Directive (85/337/EEC), which is a framework directive addressed to its Member States on the assessment of the effects of certain public and private projects on the environment. This Directive was approved in 1985 and requires each Member State to establish a procedure whereby, before developers of certain types of projects are given authorization to proceed with a project, those projects are subject to an assessment of their environmental effects.

Since Jordan is not a Member State, the Directive only has guidance status where the EIB is involved in financing projects (loans and grants) in Jordan. EIB is involved in financing irrigation and agricultural projects where EIA is required in certain cases.

The United States (USAID) EIA Guidelines

These are adapted from the main NEPA EIA regulations and are followed in preparing EIA studies for projects financed by the United States Agency for International Development (USAID). A recent example of an EIA study carried out according to these guidelines is the EIA study of the Wadi Musa Wastewater Collection and Treatment Project (CDM 1996)

Box 13.4   Environmental Assessment of the Red Sea – Dead Sea Canal – Jordan

Type of project: Hydrostatic Desalination of Sea Water (Category A).

Scope and objectives: The project involves building a 180 km canal/tunnel between the Gulf of Aqaba at the Red Sea and the Dead Sea in the Jordan Valley. The Canal will carry more than 3 billion cu.m/year of sea water to be dropped from the normal sea level to the Dead Sea level of –400 m (below normal sea level). The difference in level will be utilized to desalinate sea water. The project has two main objectives:

1. Production of 800 million cu.m./year of desalinated water to be used for municipal and industrial purposes in Jordan and Israel.
2. Transporting water to the Dead Sea to substitute the drastic losses in its water level which have mainly been caused by agricultural and industrial activities in Jordan and Israel during the last three decades.

Main environmental impacts: The project may result in some very complicated and unpredicted environmental impacts; however, the Interim Environmental Assessment of the project identified the following main impacts:

1. Contamination of groundwater due to leakage from the canal and any ponds, lagoons or other water bodies associated with it, as well as contamination resulting from seismic or terrorist disruption.
2. Re-establishment of stratification in the Dead Sea between deep fossil and less hyper-saline surface inflow waters.
3. Soil erosion and outwash deposition during construction.
4. Damage and siltation of coral reefs and associated ecology during construction of the canal intake structures.
5. Threats to wildlife and biodiversity from increased settlement and economic activity caused by the canal.
6. Threat to as yet undiscovered sites of archaeological, historical or other cultural importance, especially during construction.
7. Major socio-political gesture linking Jordan and Israel and improving ties between the two countries and social cohesion between adjacent local peoples from each country.

Proposed mitigation measures: The Interim Environmental Assessment of the project suggested the following mitigation measures:

1. Establish the largest proportion of the canal in a tunnel.
2. Careful location of the extraction works in the Aqaba area, ideally well away from the head of the Gulf and the main beaches and tourism areas.
3. Further investigation of the impact of sea water mixing with the hyper-saline Dead Sea water.
4. Concentrate desalination and power generation at a single site.

EIA Procedures followed: The World Bank EA OD 4.01.

Status of the project: Pre-feasibility studies completed. It is unlikely to be realized before year 2010 due to the great expenditure required (more than US$7 billion).

Reference: HARZA JRV Group (1997).

Box 13.5   Environmental Assessment of Adasiya Dam Project – Jordan

Type of project: Small-scale irrigation/drainage (Category B).

Scope and objectives: The project involves building a small diversion dam on the Yarmouk River near the Syrian borders. The main objectives of the project are:

1. Contribute to alleviating the water scarcity in Jordan by providing additional high quality fresh water.
2. Regulate the Yarmouk River water flows in winter to avoid damaging floods.

Main environmental impacts: The brief EIA of the project identified the following impacts:

- **Positive impacts** which include: providing additional water for irrigation and municipal and industrial purposes, improving water quality in the King Abdullah Canal, reducing soil erosion caused by winter floods (down stream) and creating new habitats for some species
- **Construction-related 'temporary' impacts** which include: the introduction of non-resident labour to the local community, landscape disfigurement, soil and water pollution, disposal of construction debris, and possible destruction of undiscovered archaeological remains
- **Permanent negative impacts** which include: inundation of wildlife habitat vegetation, possible deterioration of the water quality in the reservoir, impacts of human health and seismic threat, sedimentation/siltation, water pollution caused by activities upstream, changes in watershed hydrology, reduced silt in the downstream area, groundwater level change, effects on wildlife and loss of some habitats

Proposed mitigation measures: The brief EIA of the project suggested the following mitigation measures:

1. Lower the height of the dam to minimize the inundated land area.
2. Apply suitable land reclamation and soil stabilization measures in the upstream areas to minimize transported soil and silt behind the dam.
3. Maintain a sufficient flow in the downstream section of the river to support the existing aquatic species.

EIA procedures followed: The World Bank EA OD 4.01.
Status of the project: Construction planned to start in 1998.
Reference: DAR Al-Handasah Consultants (1997).

---

4   Problems Facing EIA Implementation in Jordan

The main problems that need to be overcome in order to achieve proper implementation of EIA in Jordan are:

1. The most obvious and challenging problem is the lack of explicit legislation concerning EIA (and the lack of legislated environmental quality standards). The ratification and approval of the Environment Protection Act has helped in introducing the concept of EIA but it is only the first step.
2. The GCEP (General Corporation for Environmental Protection) has now been established as the executive authority, with a legal mandate under the above

Act, and with an institutional structure that includes an EIA Directorate. However, the present lack of specific EIA procedures and a clear definition of the responsibilities to be exercised by the permitting authorities causes other problems. Bureaucratic routines and a weak planning system are sources of additional difficulties.

3. A third problem is the lack of awareness of, and negative attitudes towards, environmental issues in Jordan. This applies both to the general public and decision makers. Linked to this is a lack of motivation on the part of the public to become involved in development planning (partly due to ignorance of environmental issues and also to the feeling that their input would not make a difference to the decision). It is also linked to a lack of political will amongst politicians and decision makers to conserve the environment and natural resources. In summary, ensuring constructive and effective public participation in the country is a major challenge.

4. Other problems facing EIA implementation in Jordan are:
   - A shortage of suitably qualified personnel to undertake the various roles involved in EIA
   - A lack of financial resources combined with a poor economic climate within the country
   - A lack of environmental monitoring capabilities and institutions
   - Difficulties in implementing recommended mitigation measures (Al-Khoshman 1995)

5   Requirements for a Successful EIA System in Jordan

The following measures are proposed to help in remedying the shortcomings that have been identified and to do so within the spirit of Article 15 of the Environmental Protection Act, 1995:

- The EIA system should be linked to the existing national and regional permit/approval procedures
- It should be clarified whether there is a formal permitting or approval procedure to which the EIA can be linked, for each type of project for which EIA is to be required
- It should be clarified whether the permitting or approval procedure is operated at a national and/or local level and which organization(s) should serve as the decision making body
- The project proponent and the permitting authority should be separate bodies
- The same procedures should be applied whether the project proponent is in the public or private sector
- The permitting agency, in conjunction with the project proponent, should identify the stage in the project cycle at which EIA is required
- The permitting agency should request prior notification by the project proponent of his/her intention to submit a proposal
- The range of organizations and types of individuals to be consulted at different stages of the project planning should be established before scoping and should be included in the national EIA procedures (Al-Khoshman 1995)

- Clear guidelines on the contents and coverage of the EIS to be prepared and submitted by the project proponent should be included in the procedures
- The addition of a formal stage of reviewing the adequacy of the EIS before decision-making should be included in the procedures
- The requirements for the publication of the EIS, and for consultations based upon this before decision-making, should be included in the procedures
- Trained staff should be employed by the permitting/approval authorities to supervise the implementation of the EIA requirements

6   Conclusions

During the last 2–3 years, there have been some further initiatives to establish national EIA regulations and procedures in Jordan. These include:

- *The Capacity 21 Project.* This was initially proposed after the Earth Summit (RIO'92) and the United Nations Development Programme (UNDP) is the implementing agency. In Jordan the Capacity 21 project started on 1 January 1997 and has two main activities: establishing national EIA guidelines and strengthening the capacity of the General Corporation for Environment Protection (GCEP). A task force has been established to prepare draft EIA guidelines and, following their refinement they will be forwarded to a high level governmental committee with a view to finalizing and issuing them. The draft EIA by-law and guidelines are planned to be finalized in June 1999
- *The German Environmental Management Support Project.* This is one outcome of the 5-year German–Jordanian Environmental Management Strengthening Programme, which included a component relating to the establishment of national EIA procedures. In early 1999, the recommendations of the German project on EIA procedures were integrated with those proposed by the Capacity 21 project

The current implementation of EIA in Jordan is inadequate, and this is due to the lack of legal requirements for EIA and appropriate EIA procedures and guidelines. There have been some promising initiatives to establish a satisfactory EIA system in the country, but much remains to be achieved before this can be fully realized.

It is important to stress again that, to be successful, EIA legal procedures should be linked to the land-use planning and permitting systems, currently implemented in Jordan. This is needed to give sufficient power to the permitting authorities to request EIAs of proposed developments and for them to make satisfactory use of these assessments in their decisions.

References

Ahmed, A (1989) *Environmental Profile of Jordan*, Amman, Jordan
Al-Khoshman, M (1995) *The State of Environmental Impact Assessment (EIA) in Jordan and the Perspectives of Economic Valuation of the EIA Results*, Report prepared for CEDARE, Cairo, Egypt

The Ministry of Municipal and Rural Affairs & Environment (MMRA&E) and the International Union for Conservation of Nature (IUCN) (1991), *National Environment Strategy of Jordan* sponsored by the USAID, Amman, Jordan

Petersen, R, Al-Khoshman, M *et al.* (1995) *Protection of the Environment and Conservation of Nature in Jordan: Report of the GTZ Fact Finding Mission*, Eschborn, Germany

CDM Consultants (1996) *Environmental Assessment of Wadi Musa Wastewater Collection and Treatment Project*, Amman, Jordan

DAR Al-Handasah Consultants (1997) *Feasibility Study and Detailed Design of Adasiya Diversion Dam*, Amman, Jordan

HARZA JRV Group (1997) *Interim Environmental Assessment of the Red Sea–Dead Sea Canal*, Amman, Jordan

13.3   EIA IN ZIMBABWE

Shem Chaibva

1   Introduction

The Government of Zimbabwe manages natural resource use through a combination of strategic planning, legislation, policy and applied management programmes. Strategic plans provide a vision for the future and long-term direction. Policy defines and conveys the intent of an agency or organization to deliver programmes. Legislation provides the rules and regulations that guide people in their efforts to achieve and maintain healthy ecosystems and healthy communities.

At present, Zimbabwe does not have an Environmental Impact Assessment Act although preparations for one are being made. Currently, EIAs are mainly undertaken within the framework of Zimbabwe's Interim Environmental Impact Assessment Policy, which was introduced in 1994, and is administered by the Ministry of Environment and Tourism (MET).

This review first summarizes the history of EIA development in Zimbabwe. Then it describes the assessment arrangements which apply under the Interim Environmental Impact Assessment Policy. Finally, it reviews the challenges to EIA implementation which will need to be addressed for Zimbabwe to establish an EA system which works sufficiently satisfactorily in practice (Chaibva, 1996).

2   History of EIA in Zimbabwe

The first formal EIA in Zimbabwe was completed for the Darrowvale Dam in 1976. Although general and cursory in nature, it provided a list of the potential impacts of the dam. In the late 1970s and early 1980s, the Natural Resources Board commissioned two EIAs for dam projects in the Zambezi Valley in response to requests from conservation organizations. The Mupata Gorge Hydroelectric Dam assessment was completed in 1979, and in 1982 a comparative assessment of the Batoka Gorge Dam and the Mupata Dam was completed. The number of EIAs completed in the country has continued to grow, mainly on an ad hoc basis and mostly at the insistence of the funding agencies or in response to the conservation movement within the country.

Some lessons have been learned from the application of EIA on an ad hoc basis, without a clearly defined policy. The first lesson was that some EIAs were carried out late in the project cycle, when project implementation had already commenced. This was particularly the case with the Osborne Dam EIA. The effectiveness of such EIAs was therefore severely compromised.

Another limitation was that the EIA process tended to end at EIA approval. In most cases there was no monitoring programme. The only probable exception was the EIA for the Mobil Oil Exploration in the Zambezi Valley, where Departmental staff did the monitoring during project implementation in order to ensure adherence to the operational recommendations contained in the EIA.

Also, because of lack of clear policy guidelines, there was confusion as to who was supposed to do what, and how and when, in relation to environmental impact assessment. Again, due to lack of a clearly-defined Government policy on EIA, many proponents did not feel obliged to undertake EIA prior to project approval, even in cases where there was a definite need for one. The EIAs were only carried out when a decision to approve had already been made by the decision-makers.

3   The Statutory and Policy Basis for EIA

Statutory Basis

Currently, as explained above, Zimbabwe does not have an Environmental Impact Assessment Act. Existing environmental legislation is, however, applied where appropriate to new developments. Seven Acts have a bearing on air, 10 Acts have a bearing on water, and eight Acts have a bearing on soil. In addition, there is overlap between institutions; three of 13 Acts are administered by more than one ministry. Six ministries are empowered to play a role in pollution control.

The Natural Resources Act, the Mines and Minerals Act, and the Forestry Act contain limited provisions for the assessment of damage to the environment caused by development. For example, the Natural Resources Act (1981), states that:

- No large dam of which the State or any other person is the owner shall proceed to construction unless the [Natural Resources] Board has reported to the Minister on the state of the catchment area of such large dam
- No soil conservation project of the State or of any other person, which involves an estimated expenditure of one hundred thousand dollars, shall proceed to construction unless the Board has reported to the Minister on the effect the project will have on the natural resources of the area which may be affected

On the basis of these kinds of assessments, mitigation measures designed to avoid, minimize, or remediate the known and potential impacts can be developed and implemented.

Policy Basis

Zimbabwe's Interim Environmental Impact Assessment Policy was introduced in September 1994 (MET 1996). It seeks to formalize and institutionalize EIA in Zimbabwe. It is administered by the Ministry of Environment and Tourism (MET) and applies to public and private sector land and water use activities.

The Interim EIA Policy is a product of extensive consultations over a period of two years with all the EIA stakeholders in Zimbabwe who included industrialists, local authorities, Government planners, non-governmental organizations, environmental groups, permitting authorities and members of the general public.

The Interim EIA policy is meant to ensure sustainable development without impeding much-needed economic investment. It is meant to strike a balance between economic development on the one hand, and the bio-physical and socio-cultural and economic environment on the other.

The EIA policy is to be applied on a goodwill basis, for a trial period of 5–10 years during which, it is hoped, lessons will be learned about its practical applicability. Its success hinges on there being goodwill on the part of the permitting authorities, the developers, industrialists and the Government. After this interim trial period, consideration will be given to legislating for EIA in Zimbabwe.

The administration of the EIA is initially centralized in the Ministry of Environment and Tourism. Permitting authorities, such as the Zimbabwe Investment Centre, Local Authorities, the Department of Physical Planning, the National Economic Planning Commission and the commercial banks, are required to refer projects with a potential to adversely affect the environment to the Ministry of Environment for EIA screening. If an EIA is found to be necessary, then the Ministry, in liaison with the proponent, draws up the terms of reference for use by EIA consultants. The Ministry is also responsible for reviewing the EIA drafts, in conjunction with other interested parties, and recommends EIA acceptance to the Minister of Environment and Tourism.

The Policy seeks to ensure that potential and known adverse cultural, social, economic and ecological impacts, resulting from land and water use development projects, are mitigated or eliminated. The *goals* of the Policy are:

- Environmentally-responsible investment and development in Zimbabwe must be encouraged through transparent, predictable, equitable and effectively administered environmental assessment policy
- The long-term ability of natural resources to support human, plant and animal life must be maintained
- A broad diversity of plants, animals and ecosystems must be conserved
- Natural processes such as the recycling of air, water and soil nutrients must be preserved
- Irreversible environmental damage must be avoided and any environmental damage must be minimized
- The basic needs of people affected or likely to be affected by a development proposal for food, water, shelter, health and sanitation must be met

- Social, historical and cultural values of people and their communities must be conserved

The Policy is based on seven *principles*:

- EIA must enhance development, by contributing to its environmental sustainability, not inhibit it
- EIA is a means for project planning, not just for evaluation
- Identifying means for managing project impacts is an essential component of the EIA Policy
- The EIA Policy depends on the normal regulatory functions of permitting authorities to implement EIA findings
- The EIA Policy involves the participation of all government agencies with a mandated interest in the benefits and costs of a project
- The EIA Policy pays particular attention to the distribution of project costs and benefits
- Public consultation is an essential part of the EIA Policy

These principles are to be used by natural resource managers to assist them in making their decisions.

Responsibilities of EIA Participants

By design, EIA is envisaged as a participatory process involving the proponent, government agencies and the public.

The *proponent* is responsible for preparing project documents, completing the EIA, meeting management requirements resulting from EIA recommendations, including mitigation of impacts and rehabilitation. In the preparation of environmental impact statements, the proponent undertakes to inform the public about: areas of public involvement; the nature and scope of the proposed project; impacts; management techniques; decisions; monitoring programmes; rehabilitation programmes.

Government agencies have the responsibility to manage the EIA process and assist in decision-making. Their duties include:

- Establishing terms-of-reference for project assessments
- Reviewing reports including the Prospectus, EIA, and follow-up monitoring reports
- Assisting the proponent to establish a public consultation process
- Contributing to social, cultural and economic evaluation of the project

The EIA Process in Zimbabwe

The environmental impact assessment is to be completed in stages. Each EIA stage is linked to each development project stage. These typically include:

- Project idea or concept
- *Pre-feasibility*. Site selection, environmental screening, initial assessment and scoping of significant issues. An Environmental Prospectus is normally to be completed at this stage
- *Feasibility*. Detailed assessment of significant impacts; identification of mitigation needs and cost/benefit analyses. A Preliminary Environmental Impact Assessment is normally to be completed at this stage
- *Design and engineering*. Detailed design of mitigation measures. If required, a Detailed Environmental Impact Assessment is completed at this stage
- *Implementation (construction and operation)*. Implementation of mitigation measures and management strategy

There are two kinds of EIA reports. The Prospectus (completed by every proponent) is a synopsis informing the MET and other agencies that a proposed land/water use project is being proposed. It provides sufficient information to allow the Ministry to determine the need for a more detailed assessment based on screening criteria. In the event that impacts do not exist or are minimal and can be effectively managed, the proponent is not required to produce an EIA study report. In such cases, the proponent is allowed to proceed.

If it is deemed likely that the project will potentially result in significant impacts, the proponent is requested to complete an EIA according to detailed terms-of-reference approved by the MET. An EIA is a comprehensive assessment of the environmental impacts of a proposed activity, and largely based on existing information and enough field work to support the level of analysis required. It is employed to identify likely impacts, to estimate their severity and significance, and to provide recommendations designed to avoid or minimize negative impacts and to enhance potential benefits. After reviewing the EIA, the proponent may be asked to furnish more information on aspects which may not have been adequately covered by the EIA. Table 13.2 contains a selection of EIAs completed in Zimbabwe, 1995–1997.

4   Challenges to EIA Implementation in Zimbabwe

There are several possible constraints on effective EIA application in Zimbabwe. Firstly, the Zimbabwe Government has embraced a structural adjustment programme which, among other things, seeks to de-regulate and reduce bureaucratic bottlenecks in investment and project implementation. There is a danger that the interim EIA policy may be viewed in some quarters as another bureaucratic hurdle which impedes or slows down much needed economic investment.

Lack of sufficient capacity in both human and financial resources to successfully implement and administer a fully-fledged EIA programme is another handicap. However, with financial assistance from the donor community, the Zimbabwe Government is building EIA capacity in both the public and private sectors through an intensive training and re-training programme (Chaibva 1998). However, it is recognized that capacity building goes further than the training of personnel. It also

Table 13.2   *Summary of Selected EIAs Completed in Zimbabwe, 1995–1997*

Project title	Proponent	Project type
Nyaminyami Safari Lodge	Zimbabwe Development Corp.	Hotel/lodge
Blanket Mine	Kinross Holdings	Mining
Pungwe-Mutare Water Project	City of Mutare	Water
Inyala Mine	AAC	Mining
Nyajezi Dam	Manicaland Development Association	Dam
Binga Matete Bay	Trinity, Travel and Travel	Harbour/marine
Gwayi-Shangani Dam	DWR	Dam
Matetsi Safari Lodge	Zambezi Safari Lodge	Tourism
Redwing Mine	Independence Mines	Mining
Sengwa Coal Project	Rio Tinto	Mining
Kunzwi Dam	GWK Consultants	Water pipeline
Bulawayo-Matabeleland Water Supply	SWECO Consulting	Water
Bulawayo Sewerage Treatment Project	Stewart and Scott	Sewer
Batoka-Hwange Safari Lodge	Matupula Hunters	Tourism
Jumanji Enterprises Tourist Camp	Jumanji Enterprises	Tourism
Tsholotsho Tourist Lodge	Matupula Hunters	Tourism

includes strengthening institutional frameworks as well as educating decision-makers.

There is also the problem of retaining qualified and trained personnel in Government. Once staff members have been trained, there is a tendency for a number of them to leave Government ministries for the private sector or neighbouring countries.

Although there is growing environmental awareness in Zimbabwe, considerable education is still needed to convince decision-makers, in both the public and private sectors, of the advantages of EIA in promoting sustainable development. Some politicians feel that EIA should not be a priority in present conditions, given our level of industrial development.

Before moving to a legislated and mandatory EIA system, there is need to review and amend other related existing legislation. The current environmental legislation is fragmented, and it will need to be harmonized and synchronized in order to make it compatible with EIA provisions and practices. Examples of such legislation, which require review, are the Town and Country Planning Act, the Natural Resources Act, the Mines and Minerals Act and the Zimbabwe Investment Act.

There is need for decentralization of EIA administration and decision-making to local authorities, particularly in relation to smaller projects. The present centralized arrangement will have to change in order to ensure the quicker processing of EIAs. The greatest challenge facing future EIA implementation in Zimbabwe is for EIA administrators to expedite it in such a manner that it will not be seen as an impediment to economic development.

At present, the EIA policy is project specific. By its very nature, the conventional (that is, project-oriented) EIA is a self-limiting and ineffective response to current scales and rates of environmental deterioration. More proactive, integrated approaches are required – in effect, a second-generation Environmental Assessment process is needed that moves beyond an 'impact fixation' to address the root causes of unsustainable development. These causes are located in the 'upstream' phase of the decision-making cycle, in the macro-economic policies and development programmes pursued by governments. This introduces the concept of strategic environmental assessment (SEA).

SEA is a promising approach to securing that policy-making takes account of sustainability principles. By definition, the SEA approach is concerned with identifying, evaluating and mitigating the potential consequences on the environment of proposed policies, plans and programmes (Sadler 1996). The ultimate aim of the EIA programme should be to extend environmental assessment to policies, plans and programmes. However, all indications are that the EIA policy is going to remain project-oriented for some time.

References

Chaibva, S (1996) EIA in Zimbabwe – past, present and future, *EIA Newsletter* **12**: 14–15
Chaibva, S (1998) Environmental assessment training for municipalities in Sub-Saharan Africa, *EIA Newsletter* **17**: 20
Ministry of Environment and Tourism (1996) *Zimbabwe Interim EIA Policy*, MET, Harare
Sadler, B (1996) *Environmental Assessment in a Changing World*, Final Report, International Study of the Effectiveness of Environmental Assessment, Canadian Environmental Assessment Agency, Hull, Quebec

14

Environmental Assessment in Development Banks and Aid Agencies

14.1 EA PROCEDURES AND PRACTICE IN THE WORLD BANK

Colin Rees

1 Introduction

Since 1989, when the World Bank adopted Operational Directive (OD) 4.00 – Annex A: Environmental Assessment, environmental assessment (EA) has become a standard procedure for Bank-financed investment projects. The directive was amended as OD 4.01 in 1991 and was converted into Operational Policy (OP) 4.01 at the beginning of 1999 (World Bank 1999). To date, well over 1200 projects have been screened for their potential environmental impact and the Bank's experience spans most sectors, virtually all borrowing member countries and a wide array of project types.

The primary responsibility for the EA process lies with the borrower; the Bank's role is to advise borrowers throughout the process and ensure that practice and quality are consistent with EA requirements and that the process is integrated effectively into project preparation and implementation. In this context, this section reviews the Bank's EA procedures and practice and evaluates the environmental assessment experience of the Bank and its borrowing member countries (World Bank 1993, 1997a).

Environmental Assessment in Developing and Transitional Countries. Edited by N. Lee and C. George.
© 2000 John Wiley & Sons, Ltd.

2   Policy Framework for EA Procedures and Practice

OP 4.01 provides the principles and procedures for implementing the EA process. It states that the purpose of EA is to improve decision making and to ensure that the project options under consideration are environmentally sound and sustainable. Further, the OP notes that the EA is a sufficiently flexible process to allow environmental issues to be addressed in a timely and cost-effective fashion during project preparation and implementation and to help avoid costs and delays due to unanticipated environmental problems. Guidance is provided on consultation with and disclosure of information to affected groups and local NGOs. Depending upon the project, a range of instruments can be used to satisfy the Bank's EA requirements: environmental impact assessment (EIA), regional or sectoral EA, environmental audit, hazard or risk assessment, and environmental management planning.

The Bank's Environmental Assessment Sourcebook (World Bank 1991) provides comprehensive support for the application of EA to all major sectors. The first volume, dealing with policies, procedures and cross-sectoral issues, was published in 1991. Specific guidance is provided on social issues, economic analysis, strengthening local environmental management capability and institutions, financial intermediary loans and community involvement and the role of NGOs. Two other volumes, issued later in 1991 as part of the same set, address critical sectoral issues, including agriculture, transportation, urban infrastructure and industry. Since then more than 26 EA Sourcebook Updates have been published and it is anticipated that these will constitute the basis for revision of the EA Sourcebook in the near future. Additional supportive documents have also been published (World Bank 1996a,b; 1997b; 1998a,b).

3   EA and the Project Cycle

Borrowing countries have full responsibility for the design, preparation, and implementation of individual projects, but the Bank is deeply involved in each of these stages. Once a project has been identified as having a high priority and being able to contribute significantly to the economic development of the country, it undergoes intensive preparation and analysis by the borrower and the Bank to ensure that it is of sound design, is well organized, and measures up to standards of economy and efficiency for implementation.

Experience has shown that for an investment to be successful, the project must be owned by the borrower and prepared in partnership with the Bank. How task managers, both on the borrower's and the Bank's side, manage this complex assignment, develop consultative and participatory approaches and integrate the different interests involved in project development will to a great extent determine the outcome of the project.

The full integration of environmental concerns in the regular operation of a proponent agency requires an understanding of both likely substantive technical issues and project processing procedures for integrating these issues. It also requires recognition that the earlier an EA is undertaken for a proposed project and findings are integrated into project design, the better the overall project result, including its environmental planning and management. Therefore, EA is to be synchronized with the project cycle, from identification through to implementation and evaluation. The EA process followed by the Bank, and its relationship to the project cycle, is described below.

Stage 1: Screening

To decide the nature and extent of the EA to be carried out, the process begins with screening at the time a project is identified. In the screening, Bank staff determine the type, location, sensitivity and scale of the proposed project as well as the nature and magnitude of its potential environmental and social impacts, and assign the project to one of the following categories:

- *Category A* projects are those expected to have significant impacts that may be sensitive, diverse or unprecedented. These require a full EA and a field visit by an environment specialist is normally necessary
- *Category B* projects have impacts which are site-specific, few if any of them are irreversible and in most cases mitigation measures can be designed more readily than for Category A projects. These projects are submitted to a more limited EA, the nature and scope of which is determined on a case-by-case basis
- A project is classified as *Category C* if it is likely to have minimal or no adverse environmental impacts. Typical Category C projects focus on education, family planning, health, nutrition and human resource development and do not normally require any environmental assessment
- A proposed project is classified as *Category FI* if it involves investment of Bank funds through a financial intermediary in sub-projects that may result in adverse environmental impacts

EA is also required for special types of projects including sector investment lending, guarantee operations, sector adjustment lending and normally for emergency recovery projects.

Stage 2: Scoping and Development of Terms of Reference

Once a project is categorized as requiring an EA, a scoping process is undertaken to identify key issues and to assist the borrower in drafting its Terms of Reference (TOR). It is essential to identify more precisely the likely environmental impacts and to define the project's area of influence at this stage. As part of this process, information about the project and its likely environmental and social effects is disseminated to local affected communities and NGOs, followed by scheduled consultations with representatives of the same groups. The main purpose of these consultations is to focus the EA on issues of concern at the local level and take local views into account.

Stage 3: Preparing the EA Report

The main components of a full EA report are:

- *Executive summary.* This should consist of a concise discussion of the significant findings of the EA and the recommended actions to be included in the project
- *Policy, legal and administrative framework.* This is a description of the policy, legal and administrative framework within which the EA is prepared. The environmental requirements of any co-financiers should be explained
- *Project description.* This should provide a concise description of the project's geographic, ecological, social and temporal context, including any ancillary investments that may be required by the project. The need for any resettlement plan or indigenous peoples development plan should be indicated
- *Baseline data.* The baseline data are used to define the study area's dimensions and describe relevant physical, biological and socioeconomic conditions, including any changes anticipated before the project begins, and current and proposed development activities within the project area, even if not directly connected to the project. Data should be relevant to decision-making
- *Environmental impacts.* This section contains the prediction and assessment of the positive and negative impacts likely to result from the proposed project. Mitigation measures, and any residual negative impacts that cannot be mitigated, should be identified. Opportunities for environmental enhancement should be explored. The extent and quality of available data, key data gaps, and uncertainties associated with predictions should be identified/estimated. Topics that do not require further attention should be specified
- *Analysis of alternatives.* The Bank's EA OP requires the systematic comparison of the proposed investment design, site, technology and operational alternatives in terms of their potential environmental impacts and feasibility of their mitigation, capital and recurrent costs, suitability under local conditions, and institutional, training and monitoring requirements. For each alternative, the environmental costs and benefits should be quantified to the extent possible, economic values should be attached where feasible, and the basis for the selected alternative should be stated
- *Environmental management plan (EMP).* This should include the set of measures to be taken during implementation and operation to eliminate, offset, or reduce adverse environmental impacts to acceptable levels. The plan should also include the actions needed to implement these measures, e.g. mitigation, monitoring, capacity development and training. The plan should provide details on proposed work programmes and schedules to help ensure that the proposed environmental actions are in phase with construction and other project activities throughout implementation. It should consider compensatory measures if mitigation measures are not feasible or cost effective

Stage 4: EA Review and Project Appraisal

After project preparation has been completed by the borrower, the Bank reviews the proposals and undertakes a project appraisal. This is a comprehensive review of the technical, economic, financial, environmental and institutional aspects of the proposal and is conducted by Bank staff, sometimes supplemented by outside consultants. Bank staff also review the EA findings and prepare a draft Project Appraisal Document (PAD) that discusses how the borrower will address social, environmental and other issues.

Once the draft EA report is complete, the borrower submits it to the Bank for review by the project team, including the environmental specialists. If found satisfactory, the Bank team is authorized to proceed to appraisal of the project. On the Appraisal mission, Bank staff review the EA's procedural and substantive elements with the borrower, resolve any outstanding issues, assess the adequacy of the institutions responsible for environmental management in light of the EA's findings, ensure that the mitigation plan is adequately budgeted, and determine if the EA's recommendations are properly addressed in project design and economic analysis.

After the appraisal mission returns and the appraisal report is issued and reviewed, formal loan negotiations begin between the Bank and the borrower. Both sides must agree on the conditions necessary to ensure the project's success, including detailed schedules for implementation. It is important, in this context, to transfer EA findings and recommendations into appropriate language for environment-related conditions, covenants and implementation schedules in the legal agreements.

Stage 5: Project Implementation and Supervision

The borrower is responsible for implementing the project according to agreements derived from the EA process. The borrower reports on: (a) compliance with measures agreed with the Bank on the basis of the findings of the EA, including implementation of any EMP, as set out in the project documents; (b) the status of mitigating measures; and (c) the findings from monitoring progress in project implementation. The Bank supervises the implementation of environmental aspects as part of overall project supervision, using environmental specialists as necessary.

Other Features of the EA Process

Institutional capacity

When the borrower has inadequate legal or technical capacity to conduct key EA-related functions (such as review of EA, environmental monitoring, or management of mitigation measures) for a proposed project, the project should include components to strengthen that capacity.

Public consultation

Consultation with affected communities is recognized as key to identifying environmental impacts and designing mitigation measures for all A and B projects proposed for financing. For Category A projects, the Bank's policy requires consultation with affected groups and local NGOs during at least two stages of the

EA process: (1) at the scoping stage, shortly after environmental screening, and before the TORs for the EA are finalized; and (2) once a draft EA report is prepared. Consultation throughout EA preparation is required, particularly for projects that affect peoples' livelihood and for community-based projects. In projects with major social components, such as those requiring involuntary resettlement or affecting indigenous people, the consultation process should involve active public participation in the EA and project development process and the analysis of social and environmental issues should be closely linked.

Disclosure of information

For consultations between the borrower and project-affected groups and local NGOs, on all Category A and B projects proposed for financing, the borrower provides a summary of the proposed project's objectives, its description and its potential impacts. For consultation after the draft EA report is prepared, the borrower provides a summary of the EA's conclusions. In addition, for a Category A project, the borrower makes the EA report available at a public place accessible to project-affected groups and local NGOs.

Once the borrower officially provides a Category A report to the Bank, the Bank distributes the summary (in English) to the Bank's Board of Executive Directors and makes the report available through its Infoshop. Once the borrower officially provides any separate Category B EA report to the Bank, the Bank also makes it available through its InfoShop.

4   Evaluating Bank and Borrowing Member Country Environmental Assessment Experience

Since July 1992, considerable progress has been made regarding the institutional and operational aspects of EA and their links to project preparation and implementation. EA is now a firmly rooted part of the Bank's normal business activity, reducing the adverse environmental impacts of Bank financed projects and in other ways influencing their design and implementation. EA training, guidance, and best practice papers, and more importantly, increasing EA experience in the field have contributed to these results.

However certain questions persist concerning the Bank's capacity to further improve the quality and effectiveness of EAs. In particular, there are questions about how to ensure adequate supervision of EA-related measures during project implementation, especially in the light of the rapidly growing number of Category A projects that will enter the active portfolio over the next few years. In addition, there is a sharp increase in the number of projects having environmental objectives and components that will also require specialized environmental supervision.

Recent reviews of EA conducted by the Bank (World Bank 1996b; 1997a) conclude that the Bank's capacity to cope with the growing EA workload needs to be carefully monitored and in some areas strengthened. To date, the Bank's Regional Environment Units have had a key role in reviewing EAs and ensuring adequate quality and follow-up. In terms of supervision, however, they have not had defined roles and resources have been limited. The relatively recent trend internalizing some environmental functions in the Country Units is to be encouraged, but it is vital that the Bank maintains a strong environmental review capacity independent of directly operational functions. As part of the process of change in the Bank, the strengths and shortcomings of the current EA system – including the roles and responsibilities of Country Units, Regional Environment Units and the Environment Department – are under constant review and adjustments are being made. The remainder of this section reviews some *major challenges* facing the Bank in implementing EA and outlines an *action plan* for strengthening the EA process.

Major Challenges

Category B projects

During the three fiscal years (1992–1995), covered by the Second Review of World Bank Experience, a range of 70 to 100 Category B projects, which required a limited environmental analysis, were approved annually. The projects included virtually all sectors, involved different types of lending instruments, and a broad spectrum of borrowing countries. The environmental analysis carried out was equally varied. In some cases, it was detailed and extensive; in others, minimal. Such differences, which may often be appropriate in view of the wide variety of project types within the 'B' category, nevertheless suggest a need to examine further the quality and appropriateness of Category B environmental analysis. This will be a subject for the next EA review covering the period 1996–1998.

Sectoral EA

EAs that take a sector-wide perspective are becoming increasingly common. This is partly a result of a wider acceptance of sectoral EAs (SEAs) as a useful tool in sectoral investment planning, permitting the integration of environmental considerations from the outset before major project-specific decisions are made. It also reflects the Bank's increasing use of 'programmatic' loans in various sectors (roads, irrigation etc.) whereby a framework for the preparation and implementation of numerous similar sub-projects is established. Experience has shown that SEA is a more effective tool than project-specific EAs in addressing sector-wide environmental issues, including alternative investment strategies, and in bringing them to the attention of decisionmakers. By addressing these issues up front, SEA can help eliminate investment alternatives that are environmentally most damaging and reduce the information requirements for subsequent project-specific EAs, allowing them to concentrate on site-specific issues and impacts. SEAs may therefore result in improved project quality and resource savings in the area of EA.

Regional EA

To date, few regional EAs (REAs) have been prepared but there is a trend towards making more use of this instrument. The main reason for the limited use to date is probably that both the Bank and its borrowers tend to take a sector-by-sector, rather than spatial, approach to development planning. While it is still too early to fully assess the effectiveness of REA, early results are positive and appear similar to those of sectoral EAs. REA provides an opportunity for a comprehensive look at geographically defined areas such as watersheds, coastal zones or urban areas.

Private sector lending

The Bank is rapidly acquiring experience in the management of environmental issues associated with loans supporting private sector development. EA approaches, especially in connection with lending through financial intermediaries, lending in support of privatization, and loan guarantees, are being developed and refined on the basis of the Bank's own experience and through a productive exchange of information with other lending institutions. Specific environmental procedures will be developed in the new area of loan guarantees.

Action Plan for Strengthening the EA Process

The Review of World Bank Experience (World Bank 1997a) identified a number of ways in which Bank EA policies and procedures may be clarified and their implementation strengthened in Bank operations. Management has accepted the main conclusions and recommendations and is therefore committed to implementing a concrete action plan to strengthen the effectiveness of EA. In addition to the measures outlined in an action plan, a number of concrete steps have already been taken. For example, some issues, identified in the Review as requiring further guidance to staff and borrowers, have been or are being addressed through the publication of new Environmental Assessment Sourcebook Updates. Additional measures include:

Strengthening environmental supervision

The Bank is taking a number of steps to improve the quality of environmental supervision. All Category A projects now require the annual participation of an environmental or social specialist in supervision. The same applies to sensitive Category B projects. In addition, the Bank's Quality Assurance Group is undertaking assessments of the quality of supervision to enhance overall in-house accountability in this area. Other measures include: giving greater attention to key environmental aspects in the development of implementation performance and impact indicators, and when preparing supervision plans; and more use of local environmental specialists and NGOs in project monitoring.

Strengthening Bank institutional capacity

A relatively small proportion of Bank staff (including project task managers) have received EA training in recent years. Accordingly, the Bank is strengthening internal EA training through a targeted programme including basic EA training for all

Task Managers of Category A and B projects and in key sectors such as energy, agriculture and infrastructure.

Building EA capacity in borrowing countries

EA training has been conducted in a number of selected countries and the Bank's World Bank Institute (WBI) is working closely with the Environment Department and the Regional Units to develop a longer-term regional training strategy for EA. The Bank is also seeking collaboration with multilateral and bilateral agencies that provide EA training, taking advantage of existing in-country capabilities where feasible.

'Upstreaming' the EA process

The Review found that the effectiveness of EA improves when the process is initiated early, before major investment decisions are made and when there is still scope for taking a comprehensive sectoral or regional view. To sustain recent progress in making EA more effective as a planning tool, the Bank will develop mechanisms – including financial ones – to facilitate the early use of sectoral and regional EAs. Additional guidance material will be developed, along with training modules.

Public consultation and analysis of alternatives

These are areas where performance still needs to improve. EA Sourcebook Updates have been prepared and the Bank is taking steps to systematically strengthen EA Terms of Reference and to conduct training in these areas. Strengthening public participation is also under review.

Translating EA recommendations into legal agreements and contract documents

Unless actions and recommendations derived from the EA are adequately referenced in loan and credit agreements, there is no legally binding borrower commitment to implement them unless required under domestic law. The Second Review finds that this step is often a precondition for effective implementation and recommends that stronger efforts be made. There is also a need to ensure that bidding and contract documents reflect environmental actions in project legal agreements. The Bank therefore intends to pay stronger attention to this important step. Options for incorporating and reflecting EA-derived measures in bidding documents and construction contracts will also be developed. Additionally, an EA Sourcebook Update will be prepared to provide guidance on the preparation of environmental legal conditions and covenants.

In conclusion, the Bank will sustain its core EA activities and continue to implement the recommendations of the 1992 and 1995 EA Reviews. However, the Bank's EA process will also need to adapt to the rapidly changing nature of investment planning. A particular challenge lies with the use of EA as a tool to help integrate environmental considerations into proposed policy and programme formulation – for instance in designing structural adjustment or policy-based lending.

References

World Bank (1991) *Environmental Assessment Sourcebook*, Environment Department, World Bank, Washington DC (3 volumes, with subsequent updates)

World Bank (1993) *Annual Review of Environmental Assessment 1992*, World Bank, Washington DC

World Bank (1996a) *Environmental Assessment (EA) in Africa: A World Bank Commitment*, Proceedings of the Durban (South Africa) World Bank Workshop, June 25 1995 World Bank, Washington DC

World Bank (1996b) *Environmental Assessments and National Action Plans*, OED (Operations Evaluation Department) Precis, December, World Bank, Washington DC

World Bank (1997a) *The Impact of Environmental Assessment: A Review of World Bank Experience*, World Bank, Washington DC

World Bank (1997b) *Roads and the Environment: A Handbook*, World Bank, Washington DC

World Bank (1998a) *Integrating Social Concerns into Private Decision-making: A Review of Comparative Practices in Mining, Oil and Gas Sectors*, World Bank Discussion Paper Number 384, World Bank, Washington DC

World Bank (1998b) *Meaningful Consultation in Environmental Assessments*, Social Development Notes Number 39, World Bank, Washington DC

World Bank (1999) *Environmental Assessment Operational Policy, OP4.01*, World Bank, Washington DC

14.2   EA PROCEDURES AND PRACTICE IN THE ASIAN DEVELOPMENT BANK

Bindu Lohani

1   Introduction

The Asian Development Bank is a multilateral development finance institution which is engaged in promoting the economic and social progress of its developing member countries (DMCs) in the Asian and Pacific Region. The Bank is owned by the governments of 40 countries from the region and 16 countries from outside the region. Its principal functions are to: (i) make loans and equity investments for the economic and social advancement of DMCs; (ii) provide technical assistance for the preparation and execution of development projects and programmes and advisory services; (iii) promote investment of public and private capital for development purposes; and (iv) respond to requests for assistance in coordinating development policies and plans of DMCs.

It is the Bank's policy to promote environmentally sound development. Therefore it:

- Implements systematic procedures for examining the environmental impact of the Bank's projects, programmes and policies

- Encourages developing member country (DMC) governments and executing agencies to incorporate suitable environmental protection measures into project and programme design and implementation procedures, and it provides technical assistance for this purpose
- Promotes projects and programmes designed to protect, rehabilitate, and enhance the environment and quality of life in DMCs
- Trains Bank staff and DMC counterparts and disseminates documentation for guidance in environmental aspects of economic development

To ensure that Bank-funded projects, programmes, plans and policies are environmentally sustainable, projects proposed for Bank financing are categorized. On the basis of its category, the Bank recommends the level of environmental assessment – such as an IEE (initial environmental examination) or an EIA (environmental impact assessment) – required for the project. Its use of environmental assessment as a tool complements the DMCs' own formal or informal environmental assessment requirements particularly in contents and format. Many of the Bank's DMCs have formalized their environmental assessment requirements through legislation.

2   Environmental Assessment Procedures

The Bank, through its Environment Division (ENVD) under the Office of Environment and Social Development (OESD), brings to bear environmental considerations on Bank activities at different planning levels. Environmental considerations are integrated at the country, region (within a country), sector and project levels using different approaches and techniques. The Bank uses different types of procedures in conducting the environmental assessment of its various investment operations, namely: project loans; programme loans; sector loans; development finance loans; and private sector loans.

Project Loans

After consultation with Projects Department staff, ENVD categorizes projects listed in the Country Assistance Plan according to their anticipated environmental impact. Each proposed programme or project is scrutinized as to its type; location; the sensitivity, scale, nature and magnitude of its potential environmental impacts; and availability of cost-effective mitigation measures. Projects are then assigned to one of the following three categories:

- *Category A*. Projects expected to have significant adverse environmental impacts. An EIA[1] is required to address significant impacts

---

[1] A typical EIA report includes the following major elements: (i) description of the project; (ii) description of the environment; (iii) anticipated environmental impacts and mitigation measures; (iv) alternatives; (v) economic assessment; (vi) institutional requirement and environmental monitoring programme; (vii) public involvement; and (viii) conclusion. The report is prepared by the borrower and reviewed/cleared by the EIA regulatory agency of the borrowing country and the Bank.

- *Category B*. Projects judged to have some adverse environmental impacts, but of lesser degree and/or significance than those for category A projects. An IEE[1] is required to determine whether or not significant environmental impacts warranting an EIA are likely. If an EIA is not needed, the IEE is regarded as the final environmental assessment report
- *Category C*. Projects unlikely to have adverse environmental impacts. No EIA or IEE is required, although environmental implications are still reviewed

Projects under environment category A and selected projects[2] in category B are normally referred to as environmentally sensitive projects. An environmental assessment – a generic name for IEE or EIA – of the project is invariably carried out for all environmentally sensitive projects. Environmental assessment is ideally carried out simultaneously with the feasibility study of the project. Summary initial environmental examination (SIEE) and summary environmental impact assessment (SEIA) reports, highlighting the main findings of the IEE and EIA, respectively, are also prepared.

For all projects under environment category A, and for selected projects under environment category B which would benefit from external review (even if a detailed EIA is not warranted), the SIEE or SEIA prepared by the borrower is submitted to the Board at least 120 days before it considers the project. In addition to the SIEE/SEIA, the original IEE or the EIA is made available to Board members upon request. The SIEE or SEIA and the IEE or EIA, if requested, may also be made available by the Bank to locally affected groups and nongovernment organizations (NGOs) through the Director of the DMC concerned on the Board, or through the Bank's Depository Library Program, except where confidentiality rules would be violated. In such cases, information is released to those permitted by the classification to receive the document.

Programme Loans

While IEEs/SIEEs or EIAs/SEIAs do not have to be prepared for programme loans, except as stated below, environmental repercussions, if any, of the policy and institutional reforms to be introduced with the loan are examined, and appropriate covenants included in the loan documents. If an investment component is included in the programme loan and specific projects are identified as a part of the programme loan, the projects are also treated as detailed in Project Loans

Sector Loans

Under the sector lending modality, the Bank finances the capital investment needs of a given sector (a) in a specified geographical area, (b) over a specified period of

---

[1] A typical IEE includes the following major elements: (i) description of the project; (ii) description of the environment; (iii) potential environmental impacts and mitigation measures; (iv) institutional requirements and environmental monitoring programme; (v) findings and recommendations; and (vi) conclusion.

[2] These projects involve deforestation or loss of biodiversity; involuntary resettlement issues; the processing, handling and disposal of toxic and hazardous substances; or other activities that are likely to be of interest to a wide external group of persons.

time, or (c) both. Policy and institutional changes introduced as conditionalities under a sector loan may have an impact on the environment, directly or indirectly. Accordingly, policy and institutional changes proposed in a sector loan are also examined to determine their environmental implications, and appropriate environmental interventions are introduced.

To establish the broad parameters (including environmental parameters) for selecting subprojects to be financed under the loan, a few subprojects are identified and appraised prior to loan approval. Feasibility studies are prepared for such subprojects during the formulation of the sector loan, and provide the executing agency with some indication of and experience with, how feasibility studies, including IEEs or EIAs as necessary, are prepared and cost and benefit parameters established. These studies also help refine the eligibility criteria being developed for the sector loan, including specific environmental criteria and concerns that need to be carefully examined during subproject selection, design, appraisal and implementation.

For environmentally sensitive sample subprojects, the SEIA or SIEE is submitted to the Board at least 120 days before Board consideration of the sector loan to demonstrate the manner of treating environmental issues that may arise during and after subproject selection. The IEE or EIA is made available to the Board upon request. After sector loan approval, for subprojects confirmed by the Bank as environmentally sensitive, the proposal and the IEE or EIA of the subproject is forwarded to the Bank for review.

Loans Involving Financial Intermediaries

Under this category, the Bank's involvement may be in the form of credit lines or equity investment.

When the Bank's investment is in the form of equity in a financial intermediary, no subprojects to be financed by the financial intermediary are directly involved. The Bank's environmental concerns, in this case, are normally addressed either through a covenant in the line of equity agreement or through the minutes of loan negotiations. These should stipulate that the financial intermediary will prepare and adopt a policy statement that in its operations the financial intermediary will ensure that the sub-borrowers will comply with the environmental regulations and requirements of the DMC government and, if necessary, the Bank.[1] If required, the Bank may also address any need for strengthening capacity building of the financial intermediary and the relevant environmental agency to deal with the environmental issues.

Where the Bank's involvement in the financial intermediary is in the form of a credit line for sub-projects, the Bank's environmental concerns are addressed at the level of the financial institution's policies, and at the sub-project level. For sub-projects below a financial threshold, the financial institution must prepare an appropriate policy statement before the credit line is approved (see preceding

---

[1] If the government's environmental guidelines are considered inadequate, the Bank's requirements will apply in addition to those of the government. If the government's capacity to administer its own guidelines, or if necessary those of the Bank, is found inadequate, an institutional arrangement may be made, whereby, for example, a consulting firm is contracted to assist the government in this regard.

paragraph), and any need for strengthening the environment agency and/or the financial institution in the DMC is addressed. For each environmentally sensitive sub-project above the financial threshold, the Bank reviews and clears the IEE or EIA before approving the sub-project.

Private Sector Loans

Because private sector entities and implementing institutions are a diverse group and their environmental capabilities range widely, the Bank adopts a flexible procedure in dealing with private sector loans. However, the Bank's substantive requirements for environmental aspects of private sector loans are similar to those for public sector loans as described above.

3 Review of Experience

The Bank's EA arrangements were originally developed in the early 1980s and have evolved into a comprehensive system since then (ADB 1994). It also claims to have one of the best environmental monitoring information systems among the multilateral development banks. In early 1996, an evaluation was undertaken of the implementation of its environmental assessment requirements particularly of the quality of environmental assessment reports, submitted by its DMCs (ADB 1996a). The evaluation, covering about 23 EIA reports, focused on seemingly weak elements of EIAs submitted so far to the Bank, namely: analysis of alternatives, public participation and economic analysis.

Analysis of Alternatives

EIA reports submitted to the Bank showed variability in the treatment of alternatives which, in a number of cases, was relatively brief. Various project alternatives should be identified and their environmental impacts assessed to determine the project alternative with the least environmental impact and cost, and maximum environmental benefit.

Ideally an EIA report should include an analysis of impacts 'with the project' and 'without the project'. Many EIA reports submitted to the Bank included this type of information which also proved useful in classifying projects according to their strategic development objective. Project classification by development objective enables calculations of the volume and number of environmental projects approved each year in proportion to the total number of approved projects.

Discussion on project alternatives according to site, alignment, size and technology to be used is also essential. Bank experience on the inclusion of this type of analysis is variable. In some cases the search for alternative technological choices could have been improved. In the case of transport projects, alternative alignments were almost always explored before an alignment was selected. However, it was not always clear whether environmental considerations were then considered as a factor in the final choice of the alignment. Site options were investigated in most cases and, in the industry sector, environmental considerations were almost always considered in the choice of the project site.

The limited discussion of project alternatives in a number of cases can be attributed to: (i) its non-inclusion in the Borrower's environmental assessment requirements; (ii) lack of local experts available for preparing a substantial and in-depth analysis; and (iii) the preparation of the EIA in isolation from the project feasibility study.

Public Participation

Public participation should be of key importance in designing a project, exploring alternatives, identifying potential environmental impacts, designing mitigation measures and in monitoring the implementation of the environmental management programme. However, the timing, purpose and legal force of such participation vary. Public consultation could build up local ownership, engender participation, and accountability in the project throughout its life.

Public participation in Bank projects most commonly occurs during project design and project implementation and operations. A comparison of Bank approved projects prior to and proposed in 1994/1995 indicates that increasingly greater attention is being paid to public participation particularly during the implementation and operation stages (ADB 1996b). This is demonstrated particularly by the number of projects that are using the services of national and community-based NGOs.

Factors that appear to have influenced the extent of public participation include:

- *Staff working knowledge in the use of tools for public participation.* Exposure to and knowledge of viable options for undertaking public participation vary considerably. Some staff have a strong background in public participation, while for others the concept is new
- *Nature and type of project.* The type and nature of project has often influenced the level of public participation in the project. For example, the assessment of agriculture projects has included a higher level of public participation, than in the case of infrastructure projects
- *Government regulations and practices.* There is a tendency to rely largely on existing DMC government practices. From the Bank's point of view, in some cases these practices are adequate, while in many cases they fall short of the Bank's minimum requirements
- *Community leaders' commitment.* A key factor in soliciting public involvement is the leadership role played by the head of the community. A strong leader will be able to motivate his people to participate in project-related meetings or discussions, while a weak one will not be able to muster public support from his constituencies

Design stage

During this stage, the most common way in which a substantial proportion of community stakeholders have been drawn into project activities is during the Initial Social Assessment (ISA). It would appear that more attention is being given to describing the project and asking for reactions from interviewees during the ISA.

The advantage of this is that the ISA is usually based on a representative sample of communities in the project area, so that responses can be expected to represent local attitudes in general. A disadvantage is that the interviewers who conduct the ISA are often not sufficiently familiar with the project, so many questions may go unanswered or answers may not always be accurate. Additionally, the ISA is usually undertaken some time before final project design and so project components may not be fully developed.

Other public participation approaches which have been used, include:

- Workshops and seminars to inform stakeholders of the project objectives and scope, and to solicit the views of participants. These often involve higher level stakeholders, for example, government officials, but less often include community representatives from the project area
- The media, especially the press, are used occasionally either to inform the general public that the project will be undertaken or to notify the public about specific actions, such as compensation arrangements for land
- In some cases an officer from the Executing Agency or the Implementing Agency has been selected as the focal point for contacts with the public and his/her responsibility is to keep agencies and individuals informed of the progress of the project

Implementation and operation stages

There appears to be a significantly heightened recognition of the role of public participation at these stages, on the part of Bank staff. This most commonly has taken the form of support to NGOs for working with communities in the project area. Examples include:

- The Ghazi-Barotha Hydropower Project where support was provided for community development activities by a national NGO that was already active in the project area. By bringing the NGO into the project, the executing agency will be able to use the NGO's existing network to disseminate information and to gain feedback from communities
- The Biodiversity Conservation Project in Indonesia where NGOs are playing a key role to ensure that community needs are addressed during implementation and operation
- The Integrated Rural Development Project in Nepal where local user groups were formed and strengthened and made an integral part of the project. The user groups will, for example, be responsible for managing forests along new rural roads to be constructed under the project and will be involved in actual road construction and maintenance
- Video tapes are to be used in some proposed Bank projects to increase the public's awareness about how they are affected by or can contribute to the project

Though Bank projects are increasingly making use of a number of participatory approaches, the greatest need is to extend their application over a greater number of projects. This may be achieved by providing staff with appropriate information on the nature and application of potential methods according to project sector, and monitoring performance and applicability of the methods used. Employment of NGOs in enhancing project design and implementation and monitoring could be promoted where appropriate.

To streamline procedures in public participation, there is a need to draft guidelines in order to delineate the scope, who will be involved, the extent of their involvement, and when to begin such initiatives. The guidelines should be incorporated in the project design and implementation stage.

A framework for mainstreaming participatory development processes into Bank operations has been put into practice (ADB 1996c). The framework presents an overview of how participatory development processes fit into Bank operations and how these procedures will be systematically incorporated into its business practices.

Economic Analysis

In general, the economic analysis section in EIA reports includes: (a) costs and benefits of environmental impacts; (b) costs, benefits and cost effectiveness of mitigation measures; and (c) a review of the environmental impacts that have not been expressed in monetary values, expressed where possible, in quantitative terms (e.g. indicating sizes, weights etc.). It is important to stress these information requirements, at an early stage in project preparation, to the DMCs, project officers of the Bank, and project and EIA consultants.

A good number of Bank SEIAs covered in the 1996 review included adequate information on economic analysis. These included the costs of mitigation measures, a summary of the economic indicators relating to the overall project, the costs of management implementing the plan and a qualitative statement about the unquantified environmental impacts of the project. Other SEIA reports included varying amounts of information relating to the economic aspects of environmental impacts.

There has been a marked improvement in the economic analysis section of the SEIA reports compared to recent years. In 1995, not only the quantity of SEIA reports circulated to the Board increased, but the quality of the economic analysis section of these reports also improved. It is important to maintain this trend. In the future, environment specialists will have to emphasize the importance of complying with Bank guidelines and following the correct format for presenting SEIA reports to the Board.

It would be helpful for the Borrowers to use the workbook *Economic Evaluation of Environmental Impacts*, prepared for the Asian Development Bank (ADB 1996b). This will help Borrowers to perform economic analysis of environmental impacts at three stages of the project cycle: (i) identification; (ii) selection of alternatives; and (iii) economic assessment of mitigation measures.

4   State of Current Practice and Recommendations

The Bank's environmental assessment requirement continues to be a major policy instrument for reaching environmental objectives while pursuing economic development. However, its potential will be fully realized only when it is used effectively within a broader framework. To realize this, the following should be fully institutionalized:

Mainstream Environmental Assessment Findings into Project Design and Implementation

Recommended mitigating measures and environmental monitoring plans specified in the EAs should be translated into specific plans of action by the Borrowers. The Bank explicitly reflects these requirements within particular and specific environmental covenants in the Loan Agreement. Experience reveals that Borrowers are more likely to comply with environmental requirements and guidelines where there are concrete yardsticks of environmental performance supported by such agreements.

During project implementation, the following should be verified: (i) the accuracy of impact identification and assessment; and (ii) the suitability or appropriateness of measures to mitigate adverse impacts. To do these, it is essential to monitor the environmental quality within the project area.

Experience gained during project implementation can be translated into lessons learned which are then used in the design of future Bank projects and formulating strategies. The Bank should regularly review the implementation of the environmental management measures for several projects in an independent Project Review Mission to be carried out once or twice a year for a DMC. This Mission could provide an opportunity for the national environmental agency to actively participate in the review, to bring the importance of environmental aspects of the projects to the forefront, and to consolidate the Bank's own environmental review work. This practice reflects a country-focus environment review which could strengthen the rapport with national environmental agency and other related agencies, and could facilitate the environmental aspects of country programming.

Continue to Build EIA Capacity in DMCs

The Bank has adopted 'capacity building for development management' as an operating objective of its medium-term strategy. This is aimed at, among others, capacity building for protection of the environment. Partly as a result of the volume of technical assistance already provided by the Bank in the Asia Pacific Region, in environmental management capacity building, the quality of environmental assessment reports submitted to the Bank has improved considerably through the years. However, the variable quality, particularly in some DMCs, indicates a need to enhance some of the Borrowers' environmental capacities. Capacity-building is a long term process. To minimize the constraints to full implementation of environmental assessment in DMCs, the Bank's efforts in environmental institution building need to be sustained over a sufficient period of time.

Provide Bank and DMCs Access to Clean Technology Databases

To facilitate the review of environmental assessment reports, the Bank and DMCs should have access to the clean technology databases of USEPA, UNIDO, UNEP and other international organizations. These databases will assist the Borrowers and Bank Staff in evaluating the technology of projects to be adopted by the Borrower and in recommending an alternative technology, when needed, which could promote the use of cleaner technologies in the DMCs.

Institutionalize the Use of Computerized EIA in the Bank and DMCs

The use of a computerized EIA system in Asia has been developed, under RETA 5544–Development of Computer-Assisted EIA. This could facilitate the preparation of EIAs as well as the review and screening of projects by Bank staff.

Continue to Enforce the Bank's Environmental Assessment Requirements for Private Sector Projects

The Bank should not exempt or lower its environmental and social requirements for private sector projects. In the involuntary resettlement policy of the Bank, it is currently required, where applicable, that the resettlement plan should be approved prior to appraisal and the summary be made available at the Management Review Meeting. In such cases, the relevant divisions within the Bank should closely co-ordinate their activities during the environmental assessment process.

Encourage Continuing Peer Review within the Bank of Environmental Assessment Reports

Peer review of environmental assessment reports submitted to the Bank is essential. When review results in a country are pooled, it can also facilitate an objective evaluation of DMCs' technical and institutional capability. However this procedure is time consuming and since the time required to review the reports is limited, this may not be a practical option for all cases. Nevertheless, the EA reports of selected critical projects such as power development projects, mining and industry projects and other urban development with sub-projects should be reviewed in-house.

References

ADB (1994) *The Environment Program of the Asian Development Bank: Past, Present and Future*, Asian Development Bank, Manila

ADB (1996a) *Evaluation Study of the Bank's Experience in the Preparation and Review of Environmental Assessment Reports*, Asian Development Bank, Manila

ADB (1996b) *Economic Evaluation of Environmental Impacts: A Workbook*, Asian Development Bank, Manila

ADB (1996c) *Mainstreaming Participatory Development Processes*, Asian Development Bank, Manila

14.3   OECD GUIDELINES FOR BILATERAL DONORS

Remy Paris

1   Introduction

In 1989, DAC (the Development Assistance Committee of OECD) established a Working Party on Development Assistance and Environment specifically mandated to 'ensure that environmental considerations were effectively brought into policy making at the level of programme and project design and the policy dialogue with developing countries' (OECD 1991). A central focus of the Working Party has been on improving and harmonizing the environmental impact assessment practices of aid-supported projects and programmes.

Among the first priorities of the Working Party was to summarize previous recommendations of the OECD Council on environmental assessment. These were synthesized in the *DAC Guidelines on Environment and Aid No. 1: Good Practices for Environmental Impact Assessment of Development Projects* (OECD 1992a). These Guidelines – also incorporated in the *DAC Principles for Effective Aid* (OECD 1992b), which brings together the key principles for effective aid management endorsed by the DAC, the World Bank, International Monetary Fund (IMF) and the United Nations Development Programme (UNDP) – represent the consolidated advice of the DAC to its Members regarding environmental assessment. They identify the following main elements underlying good practices:

- Environmental considerations must be fully integrated into all aspects of project selection, design and implementation
- Environmental assessments must be conducted at least for the projects identified in the 1985 Council Recommendation (OECD 1985)
- Environmental assessments should address effects on human health and the ecosystem as well as social impacts (including gender specific effects and impacts on special groups), resettlement and impacts on indigenous people
- Environmental assessments should consider alternative project designs as well as mitigation and monitoring measures
- Arrangements should be made, where possible, to obtain the views of affected populations
- Off-site effects, including transboundary, delayed and cumulative effects, should be addressed; the governments of the developing countries bear the ultimate responsibility for the design of aid-assisted projects; however, donors need to ensure that environmental assessments take into account the environmental laws and regulations of partner governments as well as the donors' development co-operation policies

The Guidelines aim to ensure that: (i) all DAC donors apply similar criteria for assessments of development projects; (ii) professionals in charge of conducting assessment are aware of the consensus among the donors in regard to assessment requirements; (iii) assessments are sufficiently comparable so that they can be used by other donors; and (iv) similar criteria are used to determine 'acceptable' disturbances to ecosystems.

In October 1992, the Working Party on Development Co-operation and Environment of the DAC established a Task Force on Coherence of Environmental Assessment for International Bilateral Aid. The objective of this initiative was to recommend ways to translate the general policy and procedural framework for environmental assessment reflected in the DAC Guidelines into operational tools. This took the form of: (a) Framework Terms of Reference for the environmental assessment of development co-operation projects; and (b) detailed Guidelines for managing environmental assessments of development projects. These were published in 1996 (OECD 1996) and are reviewed below.[1]

2   Model Terms of Reference

Terms of Reference is used here to refer to a set of administrative, procedural and technical requirements which are specified in sufficient detail to ensure the completion of a full environmental assessment and the presentation of the results in a manner which meets the legal and decision-making requirements of the sponsoring donor agency(s) and the partner country. Table 14.1 summarizes the main topics to be addressed in these assessments and their basic requirements. The Framework Terms of Reference also cover the procedural and operational activities to be undertaken in each of these topics and the stages in the project cycle at which each should be carried out (see OECD 1999 for further details).

Table 14.1   *'Framework Terms of Reference' for Environmental Assessment of Development Co-operation Projects*

Topics to be addressed	Basic requirements
A. Introduction	
Background	Introduce the project and the most critical environmental issues involved
B. Context	
The problem	Identify the basic developmental issue or problem being addressed by the proposed activity, i.e. pollution, flooding, drought, erosion, energy shortage, poor health, depressed economy etc.

Continued over

[1] The Task Force, led by Canada, involved Australia, the Commission of the European Communities, Denmark, Finland, France, Germany, Japan, the Netherlands, Norway, Sweden, UK and USA. Five other organizations were also directly involved in the Project through their status as Official Observers in the Working Party: the International Institute for Environment and Development (IIED); the International Union for the Conservation of Nature and Natural Resources (IUCN); the United Nations Environment Programme (UNEP); the World Bank and the World Resources Institute (WRI).

Table 14.1   *(Continued)*

Topics to be addressed	Basic requirements
Proposed solution	Summarize the way in which the proposed activity is expected to resolve the issue, or solve or alleviate the problem, with the emphasis on sustainability
Co-operation among jurisdictions	Summarize agreements or arrangements between the donor(s) and the partner country under which the environmental assessment is being conducted
Objectives of the assessment	State clearly the objectives of the assessment and the relationship of the results to project planning, design, implementation and follow-up
C. Institutional setting	
Legal/policy base	Summarize the legal, policy and procedural bases for environmental assessment in the partner country and the donor agency
Institutional capacity	Summarize and provide an appraisal of the strengths and limitations of the partner country in the various fields of environmental protection and management
D. Alternatives	
Alternatives to the project	Evaluating the alternatives to, and within, a project is one of the most critical aspects of an environmental assessment. It is, therefore, important that the assessment analyses include a comparison of the positive and negative impacts of the proposed project with the impacts of alternatives listed below
(a) Policy interventions	Assess the potential for achieving the basic developmental objective by interventions at the policy level
(b) Other projects	Assess the potential for achieving the basic developmental objective by implementing other projects which are substantively different from the one proposed
Alternatives within the project	Evaluate potential alternatives for key aspects of the proposed project, i.e. options for siting, waste management, energy conservation and pollution control technologies
E. Institutional and public involvement	
Institutional co-operation	Show clearly how the proposed project conforms with the overall development strategy and priorities of the partner country
Public involvement	Show how affected groups and NGOs in the partner country, and interested publics in the donor country, were given the opportunity to participate in the assessment process
F. Required information and data	
Description of project	Describe the project (design life, location, layout, size, capacity, activities), its inputs (land, raw materials, energy) and outputs (products, by-products, emissions)

Table 14.1   *(Continued)*

Topics to be addressed	Basic requirements
Description of environment	Identify study boundaries and provide baseline data on relevant (as determined from scoping results) physical, ecological, economic, social, cultural and demographic conditions within those boundaries
Information quality	Assess the quality of all information, identify data gaps, and summarize the limitations placed on the assessment from such deficiencies

G. Analysis of impacts

Positive impacts	Predict how the lives of affected people will be improved and any enhancement of natural systems resulting from project implementation
Negative Impacts	
(a) Natural resources	Predict any significant reduction in the quality of air, water and soil or loss of biodiversity
(b) Human resources	Evaluate the risk of significant deterioration in the health or well-being of the affected people
(c) Relocation and compensation	Evaluate plans for involuntary relocation and describe measures to minimize the number of relocatees
(d) Cumulative impacts	Evaluate the incremental contribution to the long-term degradation of local natural and social systems
(e) Trans-boundary impacts	Evaluate the potential for neighbouring countries to be impacted and the potential effects on the global commons
(f) Impact significance	Define the meaning of the term 'significant' and assess the significance of the expected impacts

H. Mitigation and monitoring

Environment management plan	Provide a detailed plan covering mitigation of predicted impacts, management of residual effects, relocation and compensation schemes, decommissioning, and training programmes
Environment monitoring plan	Provide a comprehensive and detailed plan covering the environmental and social variables to be monitored, the location and timing of sampling and the use to be made of monitoring data

I. Conclusions and recommendations

Project decisions	Indicate the extent to which the proposed project conforms with the general principles of sustainable development.
Technical matters	Summarize the design and operational changes that are considered critical to improving the environmental acceptability of the project

Continued over

Table 14.1   *(Continued)*

Topics to be addressed	Basic requirements
Non-technical summary	Summarize, in non-technical terms, the key findings and recommendations of the assessment, including the main economic benefits, significant environmental effects and proposed mitigation
J. Annexes	
Organization	Provide information on the assessment team, the overall approach, the organization of component studies, the schedule, the budget and independent review
Report format	Follow a pre-defined outline or format as a general guide in the preparation of the environmental assessment report(s). A suggested generic format is: executive summary; project description; summary of impacts; mitigation measures; unavoidable impacts; favoured alternative; management plan; monitoring plan; technical annex

Source: OECD 1996.

3   Detailed Guidelines for Managing Environmental Assessment

Over the years, the practice of environmental assessment has become more complex and technically sophisticated. There has also been a tendency towards greater involvement of other parties in the assessment process, the so-called 'stakeholders'. Currently, environmental assessments for large projects may:

- Be undertaken within a complex multi-jurisdictional legal/regulatory framework
- Extend over a number of months, perhaps up to two years
- Involve a wide range of participants representing donor agencies, governments, proponents, technical experts, affected publics and NGOs
- Incorporate complex scientific data and sophisticated analytical methods

This growing complexity is straining the ability of officials from the donor agencies and host governments to manage the assessment process effectively. The challenge of managing environmental assessments is now as daunting as the technical complexity. Unfortunately, guidance for those responsible for managing the assessment process lags far behind the technical directions available to those who are responsible for undertaking the assessments.

The *Detailed Guidelines* which are intended as a companion working document to the *Framework Terms of Reference*, provide guidance for managing the overall assessment process. They describe the management activities appropriate to environmental assessment at each stage of the project cycle and are summarized in Table 14.2 (see OECD 1996 for further details).

Table 14.2 *Detailed Guidelines for Managing Environmental Assessment of Development Assistance Projects (Organized According to the Stages in a Generalized Project Cycle)*

Aspects of assessment	Management activities

A. Project identification stage

Communications	• Ensure that potential partners are asked to co-operate in the proposed activity as early as possible in the planning process, particularly before screening occurs • Establish the time period over which the partners prefer to be involved in the assessment process, particularly in regard to post-completion monitoring and evaluation • Maintain and circulate a current list of contact persons in the participating donor agencies and the implementing organization within the partner country • Establish mechanisms whereby officials in the donor agencies and the implementing organization within the partner country can discuss and jointly plan assessment matters in a timely and effective manner
Policy check	• Assess the proposed project against the legal requirements, policy objectives and operational priorities of the donor agencies and the partner country • In consultation with partner country officials, consider the proposed project within the context of national and regional environmental action plans, conservation strategies and state-of-environment reports • Assess the proposed project against the provisions of international agreements to which any of the parties are signatories
Screening	• Ensure that projects are screened for potential environmental impacts in a manner which meets the procedural requirements of the donors and the partner country • Establish procedures for resolving differences in the results of environmental screening, should such occur, in a manner acceptable to all of the parties involved

B. Feasibility study stage

Level of assessment	• Ensure that, if screening determines that further assessment is required, the parties agree on the need for a 'limited' or 'full' assessment
Alternatives	• Ensure early consideration of alternatives to the proposed project at the strategic level, i.e. those: (i) involving national policy decisions; (ii) having broad-scale socio-economic implications; or (iii) impacting on the achievement of long-term sustainable development (including consideration of the 'no go option')
Scoping	• Secure agreement from the donors and the representatives of the partner country government on the need for, and objectives of, a scoping process

Continued over

Table 14.2   *(Continued)*

Aspects of assessment	Management activities
	• Ensure agreement among all parties on the process of public participation to be followed in the scoping process, including: (i) defining 'public'; (ii) assigning roles and responsibilities; (iii) recording and analysing the inputs; and (iv) distributing the results
Terms of reference	• Ensure the development of a single terms of reference acceptable to all parties that defines the administrative, procedural, technical and decision-making requirements for the assessment • Determine the roles and responsibilities of the parties in regard to the management and conduct of the environmental assessment • Ensure agreement on procedures for identifying and collecting the required information and data in the partner country, and for sharing that information and data among the parties • Take whatever measures are appropriate to maximize the institutional and technical capacity of the partner country for managing the environmental assessment
Review	• Ensure agreement on procedures for reviewing progress in the environmental assessment, and for reviewing, publishing and distributing the final assessment report in donor and partner countries
C. Appraisal stage	
Project decision	• Ensure that the donors and the partner country agree on procedures to be followed after the completion of the assessment for arriving at a final decision concerning the acceptance or rejection of the environmental effects of the project • Ensure that the donors and the partner country have agreed on the environmental standards to be applied
Recommendations	• Ensure that recommended environmental management and mitigation plans, compensation schemes and monitoring are included in project approval documents
Responsibilities	• Ensure clear allocation of responsibilities for implementation of all actions contained in the assessment report
D. Final design and implementation stage	
Mitigation	• Establish procedures to ensure that recommended mitigation, monitoring, relocation and compensation plans are implemented (compliance audit)
Reporting	• Ensure that implementation reports are generated on a regular basis and distributed among the parties

Table 14.2   *(Continued)*

Aspects of assessment	Management activities
E. Operational stage	
Monitoring	• Ensure that environmental monitoring is conducted in accordance with the recommended monitoring programme, and that procedures are agreed upon in the event that limits are exceeded
	• Ensure that monitoring records are maintained and verified, and distributed to the participating donors and interested Members of the public within the partner country
Mid-term review	• Determine the parties to be involved in a mid-term environmental review, the terms of reference for the review and to whom the results will be distributed
F. Monitoring and evaluation stage	
Monitoring	• Ensure that the recommended environmental monitoring programme has been implemented, determine if it needs to be modified in light of experience, and whether the results are useful
Evaluation	• Determine if and when the parties wish to participate in an environmental evaluation of the project, and how they might co-operate given that evaluations are normally undertaken by specialized evaluation units or independent bodies.
	• Ensure agreement on the scope of the evaluation, i.e. whether it will be confined to the original terms of reference
	• Ensure agreement on the publication and distribution of the evaluation reports within the donor and partner countries
	• Based on the evaluation results, prepare a preliminary list of key environmental problems and socio-economic concerns that would likely have to be addressed at the time of project decommissioning

Source: OECD 1996.

4   Using the 'Framework Terms of Reference' and Detailed Guidelines

Given the need for terms of reference to be custom-designed for specific projects, it would be unrealistic to expect a single terms of reference to meet the needs of all donors for all projects. The *Framework Terms of Reference* are intended as a guide for developing detailed, project-specific terms of reference. It is a task for the individual donors to interpret, translate or expand the generic headings in ways that are relevant to their particular project. Although the Framework clearly reflects the expectation that the partner country will take responsibility for the preparation of the assessment, in a broader context it is meant as a common reference point for all parties involved in the assessment process.

The terms of reference for an environmental assessment can be linked to the stages in a project cycle in three time-related ways. First, they are normally applied during the pre-feasibility or feasibility stages, and the results of the ensuing assessment should be incorporated into project design and management documents. Second, some of the information required by the terms of reference may have to come from documents associated with previous stages in the project cycle, for example the project concept stage. Third, implementing the terms of reference, i.e. completing the assessment, will likely result in significant changes in later stages of the project, such as the design and monitoring stages.

The most important audience for an assessment are the senior officials charged with the ultimate responsibility for project decisions. These people need critical information on which to make such decisions, and they seldom have time to read and absorb all of the technical content of assessment documents (see OECD 1996).

The *Detailed Guidelines* are structured according to the general stages in the project cycle and according to the sequential aspects of the assessment. They are designed for application from the earliest point at which a project is considered through to final evaluation. While they are considered to be generally valid for all projects, the relevance of particular topics will depend upon the characteristics of the specific project under consideration. The 'Management Activities' are written as a set of brief instructions to those responsible for the overall management of the assessment (see Table 14.2). Additional 'Explanatory Notes' provide points of elaboration, caution or experience, but are meant to be illustrative only (see OECD 1999 for further details).

By providing a common reference point for all parties involved in the assessment process, these documents are expected to contribute to improving coherence between the assessment methodologies used by different agencies. This material is equally relevant to the needs of officials in donor agencies and their counterparts in developing countries.

References

OECD (1985) Recommendation at the Council on Environmental Assessment of Development Assistance Projects and Programmes (Adopted by the Council at its 27th Meeting), OECD, Paris

OECD (1991) *Development Co-operation, 1991 Report,* Efforts and Policies of the Members of the DAC, OECD, Paris

OECD (1992a) *DAC: Guidelines on Environment and Aid No.1, Good Practices for Environmental Impact Assessment of Development Projects,* Development Co-operation Directorate, Paris

OECD (1992b) *Development Assistance Manual, DAC Principles for Effective Aid* DAC, OECD, Paris

OECD (1996) *Coherence in Environmental Assessment: Practical Guidance on Development Co-operation Projects,* OECD, Paris, updated in 1999 and available as an electronic publication at http://www.oecd.org/dac/htm/pubs/p-coh2.htm.

15

Strengthening Future Environmental Assessment Practice: An International Perspective

Hussein Abaza

15.1 Introduction

The origins of Environmental Impact Assessment (EIA) have come back to haunt its present-day proponents. Developed in the western world in the 1970s and '80s, EIA was designed to warn of impending environmental problems, particularly pollution, from proposed projects in those countries. EIAs were applied relatively late in the project cycle, after many of the design decisions had already been made, and the environmental assessments were undertaken quite separately from the techno-economic feasibility studies (Sadler 1996). When, by the 1980s, EIAs began to be used in the context of development projects in developing countries, they were largely donor-driven and conducted by expatriate consultants with little involvement or enthusiasm on the part of the recipient countries.

More recently, environmental issues have taken centre-stage as part of the sustainable development debate and there is a growing awareness among developed and developing countries of the need to incorporate environmental concerns into the development process (Abaza 1996; Goodland and Sadler 1993; Goodland and Tillman 1995; Kirkpatrick and Lee 1997). Environmental issues are now firmly on the political agenda of many countries, as governments react to the demands of an

Environmental Assessment in Developing and Transitional Countries. Edited by N. Lee and C. George.
© 2000 John Wiley & Sons, Ltd.

increasingly environmentally conscious public (Donnelly *et al.* 1998; Chapters 2 and 3 in this volume). Yet, while these developments have certainly helped create good business for EIA practitioners, the widespread application of EIA as an effective tool for environmental management is still limited by its image as a separate, stand-alone technique to sanction project-level interventions.

On the other hand, as EIA has been around for some time, albeit as a flawed and partial tool, it has become an accepted part of project planning. As such, it offers a basic framework which can be amended and expanded to enable EIA to address not just environmental damages at the project-level, but also sustainable development objectives at the policy and programme level. This would create a very powerful tool to help policy makers consider the real trade-offs involved in operationalizing sustainable development.

Strengthening EIA would allow a balanced approach to development which takes into account environmental as well as social and economic considerations. It would provide practitioners and policy makers with the implications of the potential negative environmental, as well as social, impacts of the proposed activity and action required to avoid or mitigate these impacts to acceptable levels.

Obviously much needs to be done if this transformation is to be achieved, and this chapter sets out the main challenges and how UNEP and other international organizations can play a part in tackling them.

15.2   Key Challenges

If EIA is to become widely used as a strategic and integrated tool for environmental management and sustainable development, significant changes need to be made in:

1. The current EIA methodology.
2. The perception of EIA, particularly among developing country governments.
3. The capacities of developing countries to develop and implement their own environmental assessments.

It should be stressed that considerable efforts have already been made in all three of these areas by a range of different institutions in both developed and developing countries. The key challenge now is to build on and expand these achievements.

EIA Methodological Limitations

As it stands, much EIA work focuses almost entirely on the biophysical environment, without taking into account relevant social or economic issues. In doing so it takes care of only one of the three conditions for sustainable development, and it can actually encourage disintegration of environmental, economic and social analyses. Most often, different teams are put together to conduct separate environmental, economic and social assessments, and little effort is made to link the three pieces of work (Kirkpatrick and Lee 1997; Lee and Kirkpatrick 1999; Chapter 10 in this volume).

The fact that EIA originally evolved without consideration of the socio-economic component may seem something of a paradox, given that people are at the centre of the process of sustainable development (Principle 1 of the Rio Declaration, United Nations 1992). However, this limitation is a logical consequence of EIA's origins. In developed countries, where EIA was developed, the socio-economic situation of the affected people was not the overriding concern of the project planners nor was it perceived as being immediately and inextricably linked to environmental conditions. Obviously, these assumptions rarely hold in developing countries, where even a small scale project can have considerable social and economic impacts on the local communities. In these cases, projects may appear at first glance to offer significant positive economic impacts, but may actually result in serious social and environmental costs, all of which need to be weighed up against each other. Reconciling competing environmental, economic and social objectives in development decision-making remains the chief concern in re-designing EIA.

Project-level Focus

EIA evolved as an aid to project decision-making, not as a planning and management tool. There is, therefore, little experience in applying it in the wider context for which it is increasingly required – for sector-wide, regional and national policies, plans and programmes. Non-project activities which would benefit from EIA include: structural adjustment programmes; national budgets and development strategies; regional and sectoral activities; international trade policies; activities with transboundary environmental effects; and even global issues such as climate change and biodiversity conservation (Abaza 1996; UNEP 1996b; Sadler 1996; Kirkpatrick and Lee 1997).

The current debate on expanding EIAs into Strategic Environmental Assessments (SEAs) has generated considerable interest, but practical methods to do so are still often lacking (Canter and Sadler 1997; Lee and Walsh 1992; Sadler and Verheem 1996; Therivel and Partidario 1996; Wood and Djeddour 1992). Clearly, many of the EIA methods developed at the project level are not transferable to environmental assessments at a policy level.

Too Late in Project Cycle

Although EIAs represent a very small proportion of the total costs of a project, they do involve additional costs, and therefore tend to be commissioned at a fairly advanced stage in the project planning process. As a result, EIA is generally applied too late in the project cycle to have a real impact on the environmental issues, or indeed on any sector or dimension not already targeted by the project (Goodland and Mercier 1999).

The application of EIA and the adoption of mitigating measures to offset environmental impacts can create the illusion that the modified project represents some sort of 'optimum' project design in terms of its overall impact and use of resources. In reality, what is achieved is simply a minimization of the possible negative environmental impacts and the resulting project is usually still quite different from the project which *would have* resulted had it been initially designed to take into account environmental and social considerations. In other words, correcting for negative impacts late in the project planning stage, which is effectively what EIA often does, is not equivalent to a holistic planning approach.

EIA's impact is also limited by the fact that it rarely continues past the project planning stage into implementation and monitoring. This is partly due to the involvement of international EIA consultants whose job is finished on submission of the EIA report. They do not usually consider the need to build in follow-up and monitoring of the project's environmental performance and in any case project planners may regard this monitoring as too costly and time-consuming.

Political Resistance and Scepticism

Since EIA evolved in the western world as a means of assessing and mitigating environmental damage, it is generally perceived as essentially a negative statement, and its impact assessment techniques as reflecting a protectionist approach. One of the consequences of this negative connotation is the perception that EIA is anti-development, and a significant number of developing countries and countries in transition are reluctant to integrate EIA into their development process.

Linked with this is the sensitivity of developing countries to what they see as environmental imperialism or environmental conditionality by developed countries and donors. Tying project approval or funding to their compliance with certain environmental conditions is unacceptable to many of these countries, particularly now that they are seeing the superficiality of the commitment by some developed countries to environmental goals. Many developed countries' governments are beginning to realize the tough political and economic implications of their taking a firm stance on environmental issues. So their bold statements about their concern for the environment are not matched by any real political will to translate these intentions into actions. Some developed countries are even beginning to renege on their responsibilities to international agreements and programmes, and developing countries find the environmental requirements imposed by these countries' development agencies harder and harder to accept. From their point of view, developed countries are paying lip service to environmental issues while expecting developing countries to bear the costs of sustainable development. Several international donor agencies are also regarded as operating a system of double standards. Development projects requested by the developing countries are subjected to strict environmental assessments, while donor-initiated projects, particularly those with no clear environmental focus, are often not required to go through the same screening process.

Institutional and Capacity Constraints

EIAs in developing countries are often planned and implemented by international consultants, and opportunities to build the capacity of local institutions to perform such analyses are often wasted. And, as each donor agency generally has its own administrative and reporting requirements, with increasingly sophisticated and specialized assessment techniques, the in-country institutions often find it difficult to keep up-to-date with the necessary skills and information (OECD/DAC 1996; Donnelly *et al.* 1998).

The predominance of international consultants can cause other problems. Since EIA consultants are often on fixed-price contracts, the budgets are inflexible and do not lend themselves to the exploratory nature of EIA. In addition, EIA practitioners may not be fully independent, being hired by the project designers or contractors. This also raises the question of the appropriate level of independence for an EIA activity. Should the EIA be undertaken as a separate exercise with its team having essentially an 'auditing' role, or should it be an integral part of the overall planning and design of the project?

Low and middle income countries have capacity building requirements for environmental assessments that are very similar; and they often have policies for a lean government and budgetary constraints. Accordingly most of the new environmental work would have to be done by reorienting, training and upgrading existing professionals. Naturally there is the more general requirement that all ministries and departments should take into account the effect of their policies, plans, programmes and projects on the environment. Given the constraints of funds and professional staff this might best be achieved in the medium term by creating in-house environmental analysis in each relevant government agency, whereby environmental analysts would be team players in the division, showing through environmental assessments the wise use of resources and their incorporation into sectoral policies to make them more effective. Given the dire shortage of environmental expertise, skills and information should be introduced incrementally, first in the main natural resource ministries and departments, and then in the ministries and agencies overseeing large infrastructure projects.

15.3   The Way Forward

In considering how best to tackle the above-mentioned problems to meaningfully address sustainable development goals, it needs to be borne in mind that improving EIA methodology and practice is only part of the solution. The promotion of quality, country-driven EIAs needs to be complemented by the implementation of a series of other measures, including:

- Environmental legislation and environmental standards
- Market-based economic instruments
- The use and development of environmental and natural resource valuation methods
- Environmental awareness-raising and capacity-building
- Ample opportunities for public participation
- Appropriate institutional arrangements
- Well-developed and reliable information and data systems

Integrating Environmental and Economic Analysis

The link between EIA and the use of economic instruments is a particularly important one (World Bank 1998). Used together, they can offer decision-makers practical solutions to environmental problems, as illustrated by the following
hypothetical example. An integrated EIA of a city's five-year development plan
identified an expected increase in the amount of water pollution, resulting from the
planned construction of an industrial park on the banks of the main river. In
outlining the alternative mitigation measures possible, the EIA includes an analysis
of potential environmental and economic benefits to be gained by introducing a
wastewater effluent charge (a type of economic instrument) levied on polluting
industries, based on the amount and concentration of pollutants in their wastewater. In this way, EIA is not just a negative statement of environmental damage but a
vehicle for assessing alternative mechanisms by which the damage can be mitigated.

In order to determine the appropriate level of environmental charge or tax, to
reflect the true cost of pollution or natural resource use, valuation methods need to
be applied. Valuation methods can also be used to help establish the trade-offs
involved in reducing the quantity and/or quality of the natural resources, enabling
planners to consider the environmental costs of the project or policy concerned.

More work needs to be done on valuation methodologies to show how they can
be incorporated into EIAs (see Chapter 6 in this volume). This might entail highlighting 'win-win' scenarios in which EIA has helped to mitigate environmental
degradation while offering solutions to increase the economic performance of projects and improve the allocation of resources. More generally, tools and techniques
used for the three main disciplines of sustainable development (that is, social,
environmental and economic assessments) need to be developed in such a way that
they use a common language, in order to facilitate their simultaneous use and
eventually their full integration.

Beyond the Project Level

EIA at a project level will always be less effective in influencing the management of
natural resources while macroeconomic and sectoral policies still operate without
taking into account environmental considerations. There is therefore an argument
that project-level assessments should be replaced (or at least complemented) by
more strategic, policy-level assessments, or SEAs. There is even a case for environmental appraisal to be an integral component of the entire appraisal process, starting with the policy level and continuing to the programme and project level.
Moreover, a methodology should be developed to deal with the cumulative impacts
of projects, considering the aggregated effects of different types of projects (on, for
example, pollution and noise levels and waste generation) in a given area.

There is increasing recognition of the potential significance of moving EIA upstream in the decision making process (Goodland and Sadler 1993; Munasinghe
1996). This could ensure that environmental considerations are integrated from the
outset in the design and formulation of policies and programmes. This would not
only lead to the re-orientation of short term development activities, but to the

adoption of new development models and paths which are sustainable. Strategic environmental assessment should not be restricted to an assessment exercise but should result in the design, development and implementation of new sustainable development models. SEAs therefore involve far reaching implications and challenges which many governments may not be willing to accept. SEA involves the integration of environmental considerations in macroeconomic policies, including trade policies and agreements, structural adjustment programmes, national budgets and privatization programmes.

Co-ordination and Capacity Building

The need for co-ordination and coherence in the EIA guidelines and practices of different international agencies has been a subject of discussion for some time (Donnelly *et al.* 1998). Developing country governments and project planning consultants are confronted by a bewildering array of procedures and are given different terms of reference to meet the requirements of different agencies. Where projects are jointly funded by several agencies there is often confusion as to which set of standards applies. The project planners are typically unfamiliar with many of the procedures, so they do not understand the significance of the differences, and sometimes do not even have the appropriate guidelines. Some initiatives have been undertaken to increase co-ordination among donors and multi-lateral banks but more needs to be done, particularly in the area of capacity-building (OECD/DAC 1996).

Up until now, most EIA training has been conducted on a rather ad-hoc basis, with the different donors promoting their own individual methodologies, sending confusing signals to the developing countries concerned (UNEP 1994; Sadler 1996). The training activities did not take into account the specific needs and development priorities of the developing countries, their social and cultural backgrounds, or the level of pressure being exerted on their natural resources. Numerous EIA training manuals and guidelines have been produced for developing countries by bilaterals, multi-laterals, UN organizations and others but in most of these cases the information was not relevant to a developing country context and the written material was rarely part of a long-term EIA capacity-building programme. There is a clear need for co-ordination within the international community to develop training materials according to the specific needs of the different target groups: EIA practitioners and potential practitioners, managers, reviewers, decision-makers etc.

This will require tailoring the training courses to cater for, among other things:

- The needs and existing capacity of the host country
- Its social and cultural conditions
- The level of economic development
- Institutional, financial and human resource capabilities

This tailoring is best done by first conducting a needs assessment to determine capacity building requirements (UNEP 1994; 1996a,b). In addition to training workshops, a capacity-building strategy should include the pairing of in-country practitioners with international consultants during the implementation of EIA work, and the facilitation of information-exchange networks. Information on training institutions in the South needs to be made available and mechanisms established to co-ordinate this information at the regional and national levels. Developing countries should be encouraged to develop their own training programmes and guidelines, building on the experiences of countries in the same region, with similar conditions and priorities (Collier and Arif 1998; George 1998).

Capacity-building efforts for EIA practitioners need to go hand-in-hand with awareness raising of decision-makers about the value of using EIA and the resources and institutional arrangements required to do EIA effectively. The resistance by developing countries to using EIA is heightened by their concern about the possible constraints which EIA might place on their development, income, employment and trade opportunities and their competitiveness. This is where the new approach to EIA is particularly important, to change the assessment's image from simply 'a negative statement on environmental damage' to 'a decision-making tool to reveal the true costs and benefits of a development activity and suggest alternative mitigating measures'. Decision-makers will then come to see EIA as an aid to their making better-informed choices in formulating and designing policies and projects, rather than an imposed requirement which may limit their choices.

Public Participation

Awareness-raising efforts should also stress the need to adopt a participatory approach in EIA and indeed in the entire planning and implementation process for policies and projects (Khadka and Shrestha 1996; Chapters 7 and 9 in this volume). Public participation will need to take different forms in different countries, depending on factors such as: cultural and social considerations; local community structures; the types of communication methods available; the level of literacy among those groups likely to be affected by or interested in the EIA process; and the existence of distinct social or linguistic groups who may require the use of particular consultation methods. Effective public participation may require adaptations to the legal and institutional environment, including:

- Full advance notification of EIA decisions
- Easy access to EIA documents, background reports and data
- Opportunities to be heard, via public hearings, community meetings, or written comments
- A written record of EIA decisions, outlining the key issues and concerns raised by the community, NGO or other participants in the EIA process, and describing how these concerns have or have not been addressed in the final decisions
- Administrative or judicial review procedures in which the adequacy of the environmental review process can be tested

A wide variety of measures can be taken to encourage and support public participation, including, for example:

- Mass communication efforts involving major stakeholders
- Creating project monitoring committees including community, NGO and private sector representatives and other relevant stakeholders, who are involved in the preparation of EIAs or in the compliance monitoring of completed projects
- Promoting the role of stakeholders and affected groups
- Strengthening of NGOs to improve their ability to participate meaningfully

15.4   The Role of UNEP and Other Institutional Stakeholders

If EIA is to achieve its full potential as an instrument for sustainable development, the challenges facing the effective use of EIA need to be addressed by a network of interested international agencies, including UNEP. UNEP's role in this area is to assist countries to enhance the integration of EIAs into the planning process at the national level. Principle 17 of the Rio Declaration, produced from the 1992 United Nations Conference on Environment and Development (UNCED), declares that 'Environmental Impact Assessment, as a national instrument, shall be undertaken for proposed activities that are likely to have a significant adverse impact on the environment and are subject to a decision of a competent national authority.' Agenda 21 specifically requested UNEP to include in its work the 'further development and promotion of the widest possible use of Environmental Impact Assessment, including activities carried out under the auspices of the United Nations specialized agencies.' Thus UNEP's Governing Council set up the Economics, Trade and Environment Unit (ETEU), one of its main areas of work being the promotion of EIA capacity building, especially in developing countries and in countries in transition.

Taking its direction from the Rio Declaration, Agenda 21, and UNEP's own mandate, the ETEU reoriented its EIA programme from its conventionally limited function to make it a tool for sustainable development.

The objectives of ETEU's work on EIA are now to:

- Address the challenges in making EIA an effective and practical tool for integrating environment and development in decision-making to achieve sustainable development
- Strengthen the EIA capacity of developing countries and countries in transition, including their capability to identify EIA capacity building needs and institutional and human resource requirements, and to elaborate their national EIA legislation

UNEP sees its role primarily as a facilitator and catalyst of efforts and initiatives implemented by a wide range of institutional stakeholders, including governments, international organizations, donor agencies and national institutions in developing countries. In designing its EIA programme back in 1992, UNEP brought together 25 experts from both developed and developing countries as a working group, to advise on the priority areas, and undertook training needs assessments and capacity-building efforts in African, Asian and Latin American countries. UNEP set out to produce, in collaboration with in-country institutions and a wide range of experts, two generic documents on EIA, to help developing countries devise their own country-specific guidelines and training courses. Thus, *EIA Training Resource Manual* (UNEP 1996a) and a document on *EIA: Issues, Trends and Practice* (UNEP 1996b) were produced and launched in 1996. The main objective of the *EIA Training Resource Manual* is to enable trainers, particularly in developing countries and countries in transition, to develop tailor-made training courses for the different target groups concerned with EIA. The use of the manual will help build local capacity to develop EIA procedures and legislation, to conduct EIAS, and to monitor and evaluate the implementation of EIA. The main objective of the document on *EIA: Issues, Trends and Practice* is to enhance the capacity of countries, particularly developing countries and countries in transition, to devise suitable country-specific EIA guidelines and to address emerging issues in the use of EIA for sustainable development. The document is based on a review and comparative analysis of existing EIA practices, highlighting the key principles and common features involved.

These resource documents were subsequently trialled for their relevance for different countries. These trials took the form of workshops in Uganda, Vietnam, Hungary and Honduras and produced very useful feedback on the strengths and weaknesses of the material. The training materials and workshops have spawned a great deal of country-driven capacity building and UNEP has received positive feedback from many different national and international organizations on the two EIA resource documents including requests for:

- Practical methodologies for Strategic Environmental Assessment
- Methodologies for follow-up and monitoring in the EIA process
- Guidance on integrating socioeconomic factors into EIA
- Methodologies to support public participation in the EIA process
- Training-of-trainers programmes
- Development of indicators of sustainability for different types of development and ecosystems
- EIA guidelines for specific sectors, such as mining or hydroelectric projects

UNEP continues to try and address these needs while encouraging and supporting national institutions to play a lead role in developing their own guidelines and conducting their own training. It does this by co-ordinating with other international organizations to help formulate a common agenda for the EIA activities of national institutions and to try and avoid any conflicting or competing activities between the different bodies. UNEP has also addressed the need for awareness raising of policy makers in developing countries.

Discussion Questions

1. What are the main challenges facing the effective use of EIA and SEA to promote sustainable development, particularly in developing countries and countries in transition?

2. What kinds of roles can international organizations play in strengthening environmental assessment practice? How best can they interact with other stakeholders at different levels to ensure broad-based participation and coordination in the use of environmental assessments?
3. What are the main reasons for the scepticism and resistance, especially among developing countries, to the use of EIAs? How might these problems be overcome?

References

Abaza, H (1996) Integration of Sustainability Objectives in Structural Adjustment Programmes Using Strategic Environmental Assessment, *Project Appraisal* **11**: 217–228

Canter, L and Sadler, B (1997) *A Toolkit for Effective EIA Practice – Review of Methods and Perspectives on their Application*, A Supplementary Report of the International Study of the Effectiveness of Environmental Assessment, Environmental and Groundwater Institute, University of Oklahoma, USA; Institute of Environmental Assessment, UK; International Association for Impact Assessment

Collier, JB and Arif, S (1998) Middle East and North Africa Region, *Environment Matters at the World Bank: Annual Review* World Bank, Washington DC

Donnelly, A, Dalal-Clayton, B and Hughes, R (1998) *A Directory of Impact Assessment Guidelines: Second Edition*, International Institute for Environment and Development, London

George, C (1998) Project on environmental assessment capacity in the Mediterranean region, in *EIA Newsletter 17*, Barker, AJ, Wood, C, Jones, CE and Scott, PJ (eds), EIA Centre, University of Manchester, UK

Goodland, R and Mercier, JR (1999) The Evolution of Environmental Assessment in the World Bank: from 'Approval' to Results, *Environmental Department Paper* No.67, World Bank, Washington DC

Goodland, R and Sadler, B (1993) *The Use of Environmental Assessment in Economic Development Policy Making* World Bank, Washington, DC, USA; The Canadian Environmental Research Council, Victoria, Canada.

Goodland, R and Tillman, G (1995) *Strategic Environmental Assessment – Strengthening the Environmental Assessment Process*, World Bank, Washington, DC

Interorganizational Committee on Guidelines and Principles for Social Impact Assessment (1995) Guidelines and Principles for Social Impact Assessment, *Environment Impact Assessment Review* **15**: 11–43

Khadka, RB and Shrestha, US (1996) *Participatory Strategic Planning for Strengthening EIA Capacity*. Proceedings of a Regional Workshop, Asian Regional EIA Programme, IUCN Nepal, Kathmandu

Kirkpatrick, C, and Lee, N (eds) (1997) *Sustainable Development in a Developing World*, Edward Elgar, Cheltenham

Lee, N and Kirkpatrick, C (eds) (1999) *Sustainable Development and Integrated Appraisal in a Developing World*, Edward Elgar, Cheltenham (in press)

Lee, N and Walsh, F (1992) Strategic Environmental Assessment: An Overview, *Project Appraisal* **7**: 126–136

Munasinghe, M (ed.) (1996) *Environmental Impacts of Macroeconomic and Sectoral Policies*, International Society for Ecological Economics; World Bank; UNEP

OECD/DAC (1996) *Coherence in Environmental Assessment: Practical Guidance on Development Co-operation Projects*, OECD, Paris

Sadler, B (1996) *International Study on the Effectiveness of Environmental Assessment: Final Report*, Canadian Environmental Assessment Agency, Hull, Quebec, Canada

Sadler, B and Verheem, R (1996) *Strategic Environmental Assessment – Status, Challenges and Future Directions*, The Hague

Therivel, R and Partidario, M (1996) *The Practice of Strategic Environmental Assessment*, Earthscan, London

UNEP (1994) *An Environmental Impact Assessment Framework for Africa*, UNEP Nairobi

UNEP (1996a) *EIA Training Resource Manual*, UNEP Nairobi

UNEP (1996b) *EIA: Issues, Trends and Practice* (prepared by Bisset, R), UNEP Nairobi

Vanclay, F and Bronstein, DA (eds) (1995) *Environmental and Social Impact Assessment*, John Wiley, New York

Wood, C and Djeddour, M (1992) Strategic Environmental Assessment: EA of Policies, Plans and Programmes, *Impact Assessment Bulletin* **10**: 3

World Bank (1998) Economic Analysis and Environmental Assessment, *Environmental Assessment Sourcebook Update*, No. 23, World Bank, Washington DC

Index